Aspects of Differential Geometry V

Synthesis Lectures on Mathematics and Statistics

Editor
Steven G. Krantz, *Washington University, St. Louis*

Aspects of Differential Geometry V

Esteban Calviño-Louzao, Eduardo García-Río, Peter Gilkey, JeongHyeong Park, and Ramón Vázquez-Lorenzo

ISBN: 978-3-031-01304-1   paperback
ISBN: 978-3-031-02432-0   ebook
ISBN: 978-3-031-00278-6   hardcover

DOI 10.1007/978-3-031-02432-0

A Publication in the Springer series
SYNTHESIS LECTURES ON MATHEMATICS AND STATISTICS

Lecture #41
Series Editor: Steven G. Krantz, *Washington University, St. Louis*
Series ISSN
Print 1938-1743   Electronic 1938-1751

Aspects of Differential Geometry V

Esteban Calviño-Louzao
University of Santiago de Compostela, Spain

Eduardo García-Río
University of Santiago de Compostela, Spain

Peter Gilkey
University of Oregon, USA

JeongHyeong Park
Sungkyunkwan University, Korea

Ramón Vázquez-Lorenzo
University of Santiago de Compostela, Spain

SYNTHESIS LECTURES ON MATHEMATICS AND STATISTICS #41

ABSTRACT

Book V completes the discussion of the first four books by treating in some detail the analytic results in elliptic operator theory used previously. Chapters 16 and 17 provide a treatment of the techniques in Hilbert space, the Fourier transform, and elliptic operator theory necessary to establish the spectral decomposition theorem of a self-adjoint operator of Laplace type and to prove the Hodge Decomposition Theorem that was stated without proof in Book II. In Chapter 18, we treat the de Rham complex and the Dolbeault complex, and discuss spinors. In Chapter 19, we discuss complex geometry and establish the Kodaira Embedding Theorem.

KEYWORDS

Arzelà–Ascoli Theorem, Baire Category Theorem, Banach–Schauder Theorem, blowup, Bochner–Kodaira–Nakano identity, Bochner Vanishing Theorem, Cauchy Integral Representation Formula, Cauchy–Riemann equations, characteristic classes, Chern classes, Chern–Gauss–Bonnet Theorem, Clifford algebra, compact operator, connection, de Rham complex, Dolbeault complex, Dolbeault Lemma, Elliptic regularity, Formula of Nakano, Fourier transform, Fréchet space, Fredholm operator, Fubini–Study metric, Gårding's Inequality, Hirzebruch–Riemann–Roch Theorem, Hodge–Beltrami operator, Hodge Decomposition Theorem, Hodge–de Rham Theorem, Hodge manifolds, Hodge $\star$ operator, Kähler metric, Kähler potential, Kodaira–Akizuki–Nakano Vanishing Theorem, Kodaira Embedding Theorem, operator of Dirac type, operator of Laplace type, Parseval's inequality, Poincaré duality, Poincaré Lemma, positive line bundle, quotient topology, Rellich Lemma, Removable singularities, Riemannian manifold, Riemann period relations, Riesz–Fréchet Representation Theorem, Schwarz space, Serre Duality, sheaf cohomology, Sobolev Lemma, Sobolev spaces, spectral theory, spin complex, Stiefel–Whitney classes, Uniformization Theorem, Weitzenböck formula.

DEDICATION This book is dedicated to

Alison Carmen Celia Dae

Fernanda Hugo Junmin Junpyo

Luis Manuel Mateo Montse

Rosalía Sara Susana

Contents

Preface

This five-volume series arose out of work by the authors over a number of years both in teaching various courses and also in their research endeavors. For technical reasons, the material is divided into five books and each book is largely self-sufficient. To facilitate cross references between the books, we have numbered the chapters of Book I from 1–3, the chapters of Book II from 4–8, the chapters of Book III from 9–11, the chapters of Book IV from 12–15, and the chapters of this book (Book V), which deals with elliptic operator theory and its applications to Differential Geometry, from 16–19.

In Chapter 16, we review the basic results from functional analysis which we shall need concerning Banach and Hilbert spaces. We attempt to present the bare essentials which we think anyone working in Differential Geometry should be familiar with. Our treatment is entirely elementary and for the most part self-contained and suitable not only for first-year graduate students but also advanced undergraduates. It is also a useful summary for professional mathematicians who do not work in functional analysis.

Chapter 17 presents the basics of elliptic operator theory. This is the heart of the matter. In Theorem 5.11 of Book II, we stated the basic properties of the spectral resolution of the Hodge–Beltrami operator $\Delta_p := d_{p-1}\delta_{p-1} + \delta_p d_p$ on the space of smooth p-forms which were necessary to establish the Hodge Decomposition Theorem. In this book, we go a bit deeper into the subject to establish the spectral resolution of an arbitrary second-order self-adjoint partial differential operator of Laplace type on a smooth compact manifold without boundary. In order to keep this material accessible to our audience of advanced undergraduates and first-year graduate students, as well as researchers in the field of Differential Geometry, we develop the necessary analytic machinery from scratch. We have not assumed any familiarity with elliptic operator theory to keep the material as accessible as possible. Thus, we have chosen to return to the origins of the subject and to bypass the calculus of pseudo-differential operators and instead use the formalism of covariant derivatives provided by a connection to define the Sobolev spaces $H_k(V)$ where k is a non-negative integer. The results from functional analysis established in Chapter 16 play a central role here.

In Chapter 18, we turn from the abstract setting to the more concrete and treat the elliptic complexes which arise naturally in Differential Geometry, i.e., the de Rham complex, the spin complex, and the Dolbeault complex. We begin by introducing Clifford algebras and Clifford multiplication as this provides a convenient unifying formalism for discussing many of the operators we shall be considering. This leads naturally to a discussion of operators of Dirac type; the de Rham complex, the spin complex, and the Dolbeault complex fall naturally into this formalism. We then discuss the associated second-order operators of Laplace type. In the real setting, we establish the Weitzenböck formula and derive Poincaré duality. We then turn our attention to spinors and spin structures and establish the Lichnerowicz formula; this is the natural analogue the Weitzenböck formula in the context of spin geometry. We discuss obstructions to the existence of metrics of positive scalar curvature for spin manifolds. We present various duality and vanishing theorems in complex geometry.

We conclude in Chapter 19 by treating complex geometry and introducing, briefly, a number of topics that rely on results in analysis. We present the general Cauchy representation formula that solves the inhomogeneous Cauchy–Riemann equation. We introduce general holomorphic manifolds and treat the fundamentals of Kähler geometry. We discuss potential theory in the Kähler setting and treat Hodge geometry. We conclude by proving the Kodaira Embedding Theorem.

Our experience is that many different fields interact with Differential Geometry and that elliptic operator theory is central. The scalar curvature is a basic geometric invariant. In Section 18.3, we use an argument of Lichnerowicz, which is based on the spin complex, to obtain obstructions to the existence of metrics of positive scalar curvature; elliptic operator theory and algebraic topology play an important role in solving a problem in Differential Geometry. Another example, which will be discussed in Section 19.4, is the Kodaira Embedding Theorem which gives necessary and sufficient conditions that a compact holomorphic manifold embeds as a holomorphic submanifold of complex projective space and hence is algebraic by Chow's Theorem. The Dolbeault complex and the study of positive line bundles is central in this result. We exhibit 4-dimensional tori which admit Kähler metrics but which do not admit positive line bundles. We hope that these examples, and others, which will be treated in this book, will convince the reader that a knowledge of the basics of analysis and of elliptic operator theory is useful in Differential Geometry.

Esteban Calviño-Louzao, Eduardo García-Río, Peter Gilkey, JeongHyeong Park, and Ramón Vázquez-Lorenzo
April 2021

Acknowledgments

Carl von Clausewitz [73] states "War is fought by human beings" (English translation Matthijs Jolies). With this in mind, since mathematics is created by real people, we have provided many images of mathematicians in the five books of this series; we think having such images makes this point more explicit. The older pictures used in this book are in the public domain. We are grateful to the Archives of the Mathematisches Forschungsinstitut Oberwolfach and other sources for permitting us to use the following images from their archives; the use of these images was granted to us for the publication of these books only and their further reproduction is prohibited without their express permission. We used the following images from the Archives of the MFO:

M. F. Atiyah (115. Author: Konrad Jacobs).

M. F. Atiyah and F. Hirzebruch (7941. Author: Konrad Jacobs).

E. Beltrami (10580. Mathematische Gesellschaft (Hamburg)).

H. Cartan (7471, Author: Konrad Jacobs).

S. Chern (10026. Author: George Bergman).

P. Dolbeault (880, Author: Konrad Jacobs/Neal Koblitz).

E. Kähler (8317, Author: Konrad Jacobs).

E. Kähler (8249, Author: Dirk Ferus).

K. Kodaira (2326, Author: Konrad Jacobs).

A. Lichnerowicz (5529, Author: George Bergman).

R. Narasimhan (21745, Author: Konrad Jacobs).

L. Nirenberg (3062, Author: Konrad Jacobs).

K. Oka (3147, Author: Konrad Jacobs).

J. Peetre (19851, Author: Hans Georg Feichtinger).

L. Pontrjagin (3347, Author: Konrad Jacobs).

F. Rellich (3476, Author: Konrad Jacobs).

E. Schmidt (3682, Author: Konrad Jacobs).

L. Schwartz (3752, Author: Konrad Jacobs).

J. P. Serre (14320, Author: Renate Schmid).

I. M. Singer (12834, Author: Gert-Martin Gruel).

S. L. Sobolev (3905, Author: Konrad Jacobs).

We are grateful to the Harvard University Mathematics Department for permitting us to use a photo of R. Bott, to Gunnar Spaar for permission to use a picture of J. Peetre taken at a conference in honor of Peetre's 65th birthday, and to the University of Lund for permitting us to use a photo of L. Garding. The photo of H. Whitney was taken by his daughter, S. Thurston.

RESEARCH This work was supported by Projects PID2019-105138GB-C21 (AEI/FEDER, Spain), ED431C 2019/10, and ED431F 2020/04 (Xunta de Galicia, Spain), and by the National Research Foundation of Korea (NRF) grant funded by the Korea government (MSIT) (NRF-2019R1A2C1083957). The assistance of E. Puffini of the Krill Institute of Technology has been invaluable.

Esteban Calviño-Louzao, Eduardo García-Río, Peter Gilkey, JeongHyeong Park, and Ramón Vázquez-Lorenzo
April 2021

CHAPTER 16

Functional Analysis

We shall discuss work of the following mathematicians, among others, in Chapter 16:

C. Arzelà
(1847-1912)

G. Ascoli
(1843–1896)

R. Baire
(1874–1932)

S. Banach
(1892–1945)

V. Bunyakovsky
(1804–1889)

A.L. Cauchy
(1789–1857)

M. Fréchet
(1878–1973)

E. Fredholm
(1866–1927)

J. Gram
(1850-1916)

D. Hilbert
(1862–1943)

T. Levi–Civita
(1873–1941)

H. Minkowski
(1864–1909)

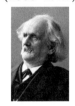
C. Neumann
(1832–1925)

J. Peetre
(1935–2019)

F. Riesz
(1880–1956)

S. Saks
(1897-1942)

J. Schauder
(1899–1943)

E. Schmidt
(1876-1959)

K. H. Schwarz
(1843–1921)

K. Weierstrass
(1815–1897)

To keep this book as self-contained as possible, we present in Section 16.1 a brief review of some elementary notions in geometry and topology. In Sections 16.2 and 16.3, we establish some standard results concerning Banach and Hilbert spaces that we will need subsequently. Section 16.4 treats the spectral theory of compact self-adjoint operators in Hilbert space.

Let $\mathbb{Z} = \{0, \pm 1, \pm 2, \ldots\}$ be the ring of integers, let $\mathbb{R}$ denote the real numbers, let $\mathbb{C}$ denote the complex numbers, let $\mathbb{F}_2$ denote the field with 2 elements, and let $\mathbb{Z}_2$ be the Abelian group with 2 elements.

16.1   BASIC CONCEPTS IN GEOMETRY AND TOPOLOGY

We begin in Section 16.1.1 by discussing the quotient topology; this is a very general construction that we use in Section 16.1.2 to topologize the coset space of an isometric action by a compact group on a compact metric space and in Section 19.2.1 to discuss complex projective space. We turn our attention to the smooth category in Section 16.1.3 where we discuss connections and in Section 16.1.4 where we discuss some notions of Riemannian geometry including the exterior derivative d and interior derivative δ. In Section 16.1.5, we introduce sheaf cohomology; we extend the discussion of Section 8.4 of Book II from the context of simple covers to the more general setting. In Section 16.1.6, we introduce the Chern, Pontrjagin, and Stiefel–Whitney classes. In Section 16.1.7, we prove the Baire Category Theorem, and in Section 16.1.8, we prove the Arzelà–Ascoli Theorem; these are fundamental results in analysis that will play a central role in our subsequent development.

16.1.1 THE QUOTIENT TOPOLOGY. Let $Y = X/\sim$ be the set of equivalence classes where $\sim$ is an equivalence relation on a compact metric space X. Let π be the canonical projection from X to Y. We define a topology on Y by requiring that $\mathcal{O}_Y$ is an open subset of Y if and only if $\pi^{-1}(\mathcal{O}_Y)$ is an open subset of X; since π is surjective and continuous, Y is compact. Furthermore, if $f_X : X \to Z$ preserves equivalence classes, then the induced map $f_Y : Y \to Z$ of equivalence classes is continuous if and only if f_X is continuous. The following is a useful observation.

Lemma 16.1   Let $f_X : X \to Z$ be continuous and surjective where X and Z are compact metric spaces. Let $x_1 \sim x_2 \Leftrightarrow f_X(x_1) = f_X(x_2)$ define an equivalence relation on X and let $Y = X/\sim$. Then the induced map $f_Y : Y \to Z$ is a homeomorphism.

Proof. Since f_Y is continuous, 1-1 and surjective, the only point at issue is the continuity of f_Y^{-1}. Let C be a closed subset of Y. Since Y is compact, C is compact. Thus, $\{f_Y^{-1}\}^{-1}(C) = f_Y(C)$ is a compact subset of a metric space and hence closed. Therefore, f_Y^{-1} is continuous.   □

16.1.2 GROUP ACTIONS. Suppose H is a compact topological group acting from the right on a compact metric space (X, d_X) by isometries. Let $x_1 \sim x_2 \Leftrightarrow x_1 = x_2 h$ for some $h \in H$ define an equivalence relation and let $Y = X/H$ be the coset space. We define a distance function on Y by setting $d_Y([x_1],[x_2]) := \min_{h \in H} d_X(x_1, x_2 h)$. Since the projection $\pi : X \to Y$ is continuous from (X, d_X) to (Y, d_Y), Lemma 16.1 shows that quotient topology on Y agrees with the metric topology on Y. The prototypical example is obtained when X is a compact Lie group and H is a closed subgroup of X and hence a Lie group as well. We then have (see Theorem 7.1 of Book II) that X/H has a natural smooth structure so that $H \to X \to X/H$ is a principal H bundle and the left action of X on X/H is smooth.

A bit of care is needed here; if H is not compact, then things can go wrong. Let $S^1 = \mathbb{R}/\mathbb{Z}$ be the unit circle. Let $H = \mathbb{Q}/\mathbb{Z}$. Then S^1/H inherits the trivial topology; the empty set and all of S^1/H are the only open subsets of S^1/H so the quotient topology is not metrizable in this instance.

16.1.3 CONNECTIONS.

We summarize briefly some material of Section 3.2 of Book I that we will need. Let $\mathbb{F} = \mathbb{R}$ or $\mathbb{F} = \mathbb{C}$. Let V be a $\mathbb{F}$-vector bundle of fiber dimension ℓ over a compact manifold M. We can find a finite open cover $\{\mathcal{O}_1, \ldots, \mathcal{O}_n\}$ of M over which V is trivial. Let $\vec{s}_i$ be local frames for V over $\mathcal{O}_i$. Over $\mathcal{O}_i \cap \mathcal{O}_j$ we may express $\vec{s}_i = \phi_{ij} \vec{s}_j$ where ϕ_{ij} is a smooth map from $\mathcal{O}_i \cap \mathcal{O}_j$ to $\mathrm{GL}(\mathbb{F}^\ell)$. The $\{\phi_{ij}\}$ satisfy the *cocycle condition* $\phi_{ij}\phi_{jk}\phi_{ki} = \mathrm{Id}$. Conversely, given such a collection, we can recover V as the disjoint union of $\mathcal{O}_i \times \mathbb{F}^\ell$ where we glue (x, v_i) to (x, v_j) using $\phi_{ij}(x)$. Let $C^\infty(V)$ denote the space of smooth sections to V; by an abuse of notation, we let $C^\infty(M)$ be the space of smooth functions on M, i.e., the space of smooth sections to the trivial bundle $\mathbb{1}$. We suppose $\mathbb{F} = \mathbb{C}$ (i.e., V is complex); the real situation is handled analogously. Since the convex combination of Hermitian inner products is again a Hermitian inner product, we can use a partition of unity to construct a smooth Hermitian inner product on the fibers of V. The pair (V, h) is then called a *unitary* or a *Hermitian* vector bundle. If the bundle in question is the tangent bundle TM, then the fiber metric is denoted by g and the resulting structure (M, g) is called a *Riemannian manifold*.

A first-order partial differential operator $\nabla : C^\infty(V) \to C^\infty(T^*M \otimes V)$ is called a connection if it satisfies the *Leibnitz formula*

$$\nabla(fs) = df \otimes s + f \nabla s \quad \text{for} \quad f \in C^\infty(M) \quad \text{and} \quad s \in C^\infty(V).$$

Connections always exist locally. Since the convex combination of connections is again a connection, we can construct connections using a partition of unity. We set $\nabla_X s = \langle \nabla s, X \rangle$ where $\langle \cdot, \cdot \rangle$ denotes the natural pairing between the tangent bundle TM and the cotangent bundle T^*M. If h is a Hermitian fiber metric on V, then the identity

$$h(\nabla_X s_1, s_2) + h(s_1, \nabla_X^* s_2) = X h(s_1, x_2)$$

defines the dual connection ∇^*; ∇ is said to be a *unitary connection* if $\nabla = \nabla^*$. If ∇ is an arbitrary connection, then we may set $\tilde{\nabla} := \frac{1}{2}(\nabla + \nabla^*)$ to construct a unitary connection.

16.1.4 RIEMANNIAN MANIFOLDS.

We summarize some material from Sections 3.1 and 3.3 of Book I. Let $\vec{x} = (x^1, \ldots, x^m)$ be a system of local coordinates on a Riemannian manifold

(M, g) of dimension m. Set $\partial_{x^i} = \frac{\partial}{\partial x^i}$. Adopt the *Einstein convention* and sum over repeated indices to express the *Riemannian metric* in the form

$$g = g_{ij}\,dx^i \otimes dx^j \quad \text{where} \quad g_{ij} := g(\partial_{x^i}, \partial_{x^j}).$$

If g^{ij} is the inverse matrix, then the dual metric on the cotangent bundle is $g^{ij}\partial_{x^i} \otimes \partial_{x^j}$. The Riemannian measure is given by

$$|\,\mathrm{dvol}\,| = \sqrt{\det(g_{ij})}\, dx^1 \cdots dx^m. \tag{16.1.a}$$

We use the notation $|\,\mathrm{dvol}\,|$ to emphasize that this is a measure rather than an m-form.

The Levi–Civita connection (see the seminal paper of Levi–Civita [46]) is the unique torsion-free connection on the tangent bundle which is Riemannian, i.e., satisfies

$$\nabla_X Y - \nabla_Y X - \nabla_{[X,Y]} = 0 \quad \text{and} \quad Xg(Y, Z) = g(\nabla_X Y, Z) + g(Y, \nabla_X Z).$$

Extend ∇ to covariantly differentiate tensors of all types. For example, we have that $\nabla g = 0$ and $\nabla|\,\mathrm{dvol}\,| = 0$. Expand $\nabla_{\partial_{x^i}}\partial_{x^j} = \Gamma_{ij}{}^k \partial_{x^k}$ to define the *Christoffel symbols*.

Lemma 16.2   Let (M, g) be a Riemannian manifold.

1. $\Gamma_{ij}{}^k = \frac{1}{2}g^{k\ell}\{\partial_{x^i}g_{j\ell} + \partial_{x^j}g_{i\ell} - \partial_{x^\ell}g_{ij}\}$.
2. Fix a point P of M. There exists a coordinate system which is centered at P so
   (a) $g_{ij} = \delta_{ij} + O(\|x\|^2)$.
   (b) $R_{ijk\ell}(P) = \frac{1}{2}\{g_{j\ell/ik} + g_{ik/j\ell} - g_{i\ell/jk} - g_{jk/i\ell}\}(P)$.

Proof. Assertion 1 is the *Koszul formula*. Assertion 2a follows from the discussion of geodesic coordinates given in Theorem 3.7 of Book I. We use Assertion 1 to compute:

$$
\begin{aligned}
R_{ijk\ell} &= g_{\ell n}(\partial_{x^i}\Gamma_{jk}{}^n + \Gamma_{ip}{}^n\Gamma_{jk}{}^p - \partial_{x^j}\Gamma_{ik}{}^n - \Gamma_{jp}{}^n\Gamma_{ik}{}^p) \\
&= \partial_{x^i}(g_{\ell n}\Gamma_{jk}{}^n) - \partial_{x^j}(g_{\ell n}\Gamma_{ik}{}^n) + O(\|x\|) \\
&= (\partial_{x^i}\Gamma_{jk\ell} - \partial_{x^j}\Gamma_{ik\ell}) + O(\|x\|) \\
&= \tfrac{1}{2}\{\partial_{x^i}\partial_{x^k}g_{j\ell} + \partial_{x^i}\partial_{x^j}g_{k\ell} - \partial_{x^i}\partial_{x^\ell}g_{jk}\} \\
&\quad + \tfrac{1}{2}\{-\partial_{x^j}\partial_{x^k}g_{i\ell} - \partial_{x^j}\partial_{x^i}g_{k\ell} + \partial_{x^j}\partial_{x^\ell}g_{ik}\} + O(\|x\|).
\end{aligned}
$$

We then simplify this expression to complete the proof.   □

Let (M, g) be a compact Riemannian manifold. Let $\exp_P : T_P M \to M$ be the exponential map. It is characterized by the property that the curves $\sigma_\xi(t) := \exp_P(t\xi)$ are geodesics in M with $\sigma_\xi(0) = P$ and $\dot\sigma_\xi(0) = \xi$. If $\{e_1, \ldots, e_m\}$ is an orthonormal basis for $T_P M$, the map $(x^1, \ldots, x^m) \to \exp_P(x^1 e_1 + \cdots + x^m e_m)$ gives a system of local coordinates on M for $\|x\|^2 < \varepsilon$ if ε is sufficiently small that are called *geodesic coordinates*. We say that an open subset of M is *geodesically convex* if given any two points of the set, they can be joined by a unique length minimizing geodesic lying entirely in the open set. The following observation follows from Lemma 3.14 of Book I.

Lemma 16.3   Let (M, g) be a compact Riemannian manifold. There exists $\varepsilon > 0$ so that the geodesic ball of radius $r < \varepsilon$ about any point P of M is geodesically convex.

The exterior derivative d maps $C^\infty(\Lambda^p M)$ to $C^\infty(\Lambda^{p+1} M)$ (see Section 2.3.3 of Book I). The *exterior coderivative* δ is the formal adjoint of d with respect to the L^2 inner product, and maps $C^\infty(\Lambda^p M)$ to $C^\infty(\Lambda^{p-1} M)$. The Levi–Civita connection defines a first-order partial differential operator

$$\nabla : C^\infty(\Lambda^p M) \to C^\infty(T^* M \otimes \Lambda^p M).$$

Let $\mathrm{ext} : T^* M \otimes \Lambda^p M \to \Lambda^{p+1} M$ and $\mathrm{int} : T^* M \otimes \Lambda^p M \to \Lambda^{p-1} M$ be *exterior multiplication* and the dual map *interior multiplication*, respectively. We obtain invariantly defined differential operators

$$\mathrm{ext} \circ \nabla : C^\infty(\Lambda^p M) \to C^\infty(\Lambda^{p+1} M) \quad \text{and} \quad \mathrm{int} \circ \nabla : C^\infty(\Lambda^p M) \to C^\infty(\Lambda^{p-1} M).$$

The following result is a restatement of Assertion 1 of Theorem 5.14 of Book II.

Lemma 16.4   If ∇ is the Levi–Civita connection, then $d = \mathrm{ext} \circ \nabla$ and $\delta = -\mathrm{int} \circ \nabla$.

16.1.5 SHEAF COHOMOLOGY. We shall present a brief introduction to the subject and refer to Griffiths and Harris [31], Grothendieck [33], and Hartshorne [37] for further details. We introduced *sheaf cohomology* in Section 8.4 of Book II to establish the de Rham isomorphism. An essential ingredient of the proof was the existence of a *simple cover*, i.e., a finite open cover of M by small geodesic balls so that the arbitrary intersection of these geodesic balls was again geodesically convex and hence contractible. Unfortunately, in our discussion of the Kodaira Embedding Theorem, we cannot restrict to a simple cover and must deal with the general setting.

Definition 16.5   Let M be a compact manifold.

1. A *pre-sheaf* S of Abelian groups over M is an assignment to each open set $\mathcal{O}$ of M an Abelian group $S(\mathcal{O})$ together with restriction maps $\varrho_{\mathcal{U},\mathcal{O}} : S(\mathcal{O}) \to S(\mathcal{U})$ which are group homomorphisms whenever $\mathcal{U}$ is a subset of $\mathcal{O}$. We require the restriction maps to satisfy $\varrho_{\mathcal{O},\mathcal{O}} = \mathrm{Id}$, and $\varrho_{\mathcal{V},\mathcal{O}} = \varrho_{\mathcal{V},\mathcal{U}}\varrho_{\mathcal{U},\mathcal{O}}$ whenever $\mathcal{V} \subset \mathcal{U} \subset \mathcal{O}$. We also require that the group associated to the empty set is the trivial Abelian group $\{0\}$ consisting of a single element.

2. We say that a pre-sheaf S is *complete* if whenever $\mathcal{O}_i$ is an open cover of $\mathcal{O}$ and whenever given elements $f_i \in S(\mathcal{O}_i)$ satisfying $\varrho_{\mathcal{O}_i \cap \mathcal{O}_j, \mathcal{O}_j} f_j = \varrho_{\mathcal{O}_i \cap \mathcal{O}_j, \mathcal{O}_i} f_i$, then there is a unique element $f \in S(\mathcal{O})$ such that $\varrho_{\mathcal{O}_i, \mathcal{O}} f = f_i$. The resulting element f is called the *extension*.

3. A complete pre-sheaf S is called a *sheaf*. Any pre-sheaf can be completed to a sheaf.

Example 16.6   One has the following examples of sheafs.

1. Let V be a smooth vector bundle over M. Let $C^k(V)$ be the sheaf which assigns to an open set $\mathcal{O}$ the space of C^k sections to V over $\mathcal{O}$. Let $P \in M$. The *sky-scraper sheaf* V_P assigns to each open set $\mathcal{O}$ which contains P the fiber of V at P and otherwise assigns $\{0\}$.

2. Let G be an Abelian group given the discrete topology. Let $\mathbb{G}$ be the *constant sheaf* which assigns to each non-empty open subset $\mathcal{O}$ the group G.

3. Let $\ker(d_p)$ be the sheaf of closed differential p-forms which assigns to every open set $\mathcal{O}$ the vector space $\ker\{d : C^\infty(\wedge^p(\mathcal{O})) \to C^\infty(\wedge^{p+1}(\mathcal{O}))\}$. Note that the assignment of the vector space $\mathrm{range}\{d : C^\infty(\wedge^{p-1}(\mathcal{O})) \to C^\infty(\wedge^p(\mathcal{O}))\}$ to $\mathcal{O}$ is not a sheaf. It is a pre-sheaf whose completion is the sheaf $\ker(d_p)$ of closed p-forms.

Definition 16.7   Let $\mathcal{U} := \{\mathcal{U}_i\}_{i \in F}$ be a finite open cover of a compact manifold M. Fix an ordering on the index set F. Let S be a sheaf over M. An element of $C^p(\mathcal{U}, S)$ is a collection of elements $f(i_0, \ldots, i_p) \in S(U_{i_0} \cap \cdots \cap U_{i_p})$. Define $\delta : C^p(\mathcal{U}, S) \to C^{p+1}(\mathcal{U}, S)$ by

$$\delta(f)(i_0, \ldots, i_p) = \sum_{j=0}^{p} (-1)^j \varrho_j(f(i_0, \ldots, i_{j-1}, i_{j+1}, \ldots i_p))$$

where ϱ_j is the restriction map from the group

$$S(U_{i_0} \cap \cdots \cap U_{i_{j-1}} \cap U_{i_{j+1}} \cap \cdots \cap U_{i_p})$$

to the group

$$S(U_{i_0} \cap U_{i_{j-1}} \cap U_{i_j} \cap U_{i_{j+1}} \cap \cdots \cap U_{i_p}).$$

One has $\delta^2 = 0$ and $H^p(\mathcal{U}, S)$ is the cohomology of the resulting chain complex. One can partially order the open covers by refinement; the sheaf cohomology groups $H^p(M, S)$ are defined as the direct limit of the cohomology groups $H^p(\mathcal{U}, S)$ over all open covers.

Theorem 8.14 of Book II shows that $H^p(M; \mathbb{G})$ is the usual topological cohomology group with coefficients in the Abelian group $\mathbb{G}$. It is immediate from the definition that $H^0(M, S) = S(M)$ if S is a sheaf. If $S_1 \to S_2$ is a morphism of sheafs, then we have a natural morphism $H^p(M, S_1) \to H^p(M, S_2)$.

Definition 16.8   The *stalk* of a sheaf S at P is defined to be $\cap_{P \in \mathcal{O}} \varrho S(\mathcal{O})$ taken over the appropriate restriction maps. We say that a short exact sequence of sheafs is exact, if the corresponding sequence is exact on the stalk level.

For example, the Poincaré Lemma and the Dolbeault Lemma (see Theorem 19.7) show that we have two short exact sequences of sheafs (see the discussion in Sections 18.2 and 18.4

for a precise definition of d_p and $\partial_{p,q}$, respectively)

$$0 \to \ker(d_p) \to C^\infty(\Lambda^p) \to \ker(d_{p+1}) \to 0,$$
$$0 \to \ker(\bar{\partial}_{p,q}) \to C^\infty(\Lambda^{p,q}) \to \ker(\bar{\partial}_{p,q+1}) \to 0. \qquad (16.1.b)$$

The following result is due to Cartan and Eilenberg [14]. Since it is proved using the same argument used to verify Lemma 8.3 of Book II, we omit the proof.

Lemma 16.9   If $0 \to S_1 \xrightarrow{i} S_2 \xrightarrow{\pi} S_3 \to 0$ is a short exact sequence of sheafs, then we have a long exact sequence in cohomology

$$\dots H^p(M, S_1) \xrightarrow{i_*} H^p(M, S_2) \xrightarrow{\pi_*} H^p(M, S_3) \xrightarrow{c} H^{p+1}(M, S_1) \xrightarrow{i_*} H^{p+1}(M, S_2) \dots$$

where c is the connecting homomorphism.

The associated long exact sequences in cohomology arising from the short exact sequences of Equation (16.1.b) are an essential ingredient in proving a theorem of de Rham (see Theorem 18.9) which identifies $H^p(M, \mathbb{C})$ with $\ker\{\Delta_p\}$ in the real setting and a theorem of Dolbeault (see Theorem 18.31) which identifies $H^q(M, \mathbf{Hol}(\Lambda^{p,0}))$ with $\ker\{\Delta_{p,q}\}$ in the complex setting. Lemma 16.9 will also play a central role in our proof of the Kodaira Embedding Theorem in Section 19.4.

16.1.6 CHARACTERISTIC CLASSES. Let $\mathbb{F} \in \{\mathbb{F}_2, \mathbb{R}, \mathbb{C}\}$. Let M be a compact manifold and let $\{\mathcal{O}_1, \dots, \mathcal{O}_n\}$ be an open cover of M by a finite number of sufficiently small geodesic balls so that the $\mathcal{O}_i$ are geodesically convex. This ensures that $\mathcal{O}_{i_1} \cap \dots \cap \mathcal{O}_{i_p}$ is either empty or geodesically convex and hence contractible; thus, $\{\mathcal{O}_i\}$ is a simple cover. Let $H^p(M, \mathbb{F})$ be the cohomology of M with coefficients in the constant sheaf $\mathbb{F}$ where

$$\mathbb{F}(\mathcal{O}) = \left\{ \begin{array}{ll} \mathbb{F} & \text{if } \mathcal{O} \neq \emptyset \\ \{0\} & \text{if } \mathcal{O} = \emptyset \end{array} \right\}.$$

$H^p(M, \mathbb{F})$ is an $\mathbb{F}$ vector space. The ring structure on $\mathbb{F}$ makes $H^*(M, \mathbb{F}) = \oplus_p H^p(M, \mathbb{F})$ into a graded ring.

16.1.6.a The Stiefel–Whitney Classes [69, 77]. Let V be a real vector bundle over M of dimension ℓ. The Stiefel–Whitney classes $\mathrm{sw}_k(V)$ are defined for $0 \leq k \leq \ell$; we set $\mathrm{sw}_k(V) = 0$ for $k > \ell$. Let

$$\mathrm{sw}(V) = 1 + \mathrm{sw}_1(V) + \dots + \mathrm{sw}_\ell(V)$$

be the *total Stiefel–Whitney class*. We refer to Milnor and Stasheff [50] for the proof of the following result as it is beyond the scope and focus of this book; these axiomatic properties characterize the Stiefel–Whitney classes.

Lemma 16.10   Let V be a real vector bundle over a compact manifold M.

1. $\mathrm{sw}_k(V) \in H^k(M, \mathbb{F}_2)$.
2. $\mathrm{sw}(V \oplus W) = \mathrm{sw}(V)\,\mathrm{sw}(W)$.
3. If $f : M \to N$, then $\mathrm{sw}(f^*V) = f^*\,\mathrm{sw}(V)$.
4. If L is the Möbius line bundle over S^1 then $\mathrm{sw}_1(L)$ generates $H^1(S^1, \mathbb{F}_2) = \mathbb{F}_2$.

The classes sw_1 and sw_2 will play a particular role in our development. Subsequently, in Section 18.3.3, we will give a very explicit description of the first Stiefel–Whitney class and show that if V is a real vector bundle, then $\mathrm{sw}_1(V)$ vanishes if and only if V is orientable. In Section 18.3.4, we will also give a very explicit construction of the second Stiefel–Whitney class and show that if V is orientable, then $\mathrm{sw}_2(V) = 0$ if and only if V admits a spin structure.

16.1.6.b The Chern Classes. We present a brief introduction to *Chern–Weil* theory and refer to Chapter 2 of Gilkey [26] for further details; there are, of course, many excellent references for this material. Let V be a smooth vector bundle of fiber dimension ℓ over a smooth manifold M. Let $\mathcal{R}_\nabla(X, Y) := \nabla_X \nabla_Y - \nabla_Y \nabla_X - \nabla_{[X,Y]}$ be the curvature operator of a connection ∇. We set

$$R_\nabla = \tfrac{1}{2} R(\partial_{x^i}, \partial_{x^j}) dx^i \wedge dx^j \in C^\infty(\Lambda^2 M \otimes \mathrm{End}(V))\,.$$

Let P be a polynomial map from the space of $\ell \times \ell$ matrices $M_\ell(\mathbb{C})$ to the complex numbers which is homogeneous of degree k. We say that P is invariant if $P(gAg^{-1}) = P(A)$ for any $g \in \mathrm{GL}(\ell, \mathbb{C})$. If P is invariant, then $P\left(\frac{\sqrt{-1}}{2\pi} R_\nabla\right)$ is a well defined element of $\Lambda^{2k}(M, \mathbb{C})$; the normalizing constant of $\frac{\sqrt{-1}}{2\pi}$ is present to simplify certain later formulas. The de Rham Theorem (see Theorem 18.9) identifies $H^p(M, \mathbb{C})$ with the de Rham cohomology groups

$$H^p_{\mathrm{deR}}(M) := \frac{\ker\{d : C^\infty(\Lambda^p M) \to C^\infty(\Lambda^{p+1} M)\}}{\mathrm{Image}\{d : C^\infty(\Lambda^{p-1} M) \to C^\infty(\Lambda^p M)\}}\,.$$

One can show $dP(R_\nabla) = 0$ and $P(V) := [P(R_\nabla)] \in H^{2k}(M, \mathbb{C})$ is independent of ∇. Let

$$c(A) := \det(I + A) = 1 + c_1(A) + \cdots + c_\ell(A)$$

define invariant homogeneous polynomials $c_i(A)$. Let $\vec{\lambda} := \{\lambda_1, \ldots, \lambda_\ell\}$ be the eigenvalues of A where each eigenvalue is repeated according to multiplicity;

$$c_i(A) = \sum_{1 \le \nu_1 < \cdots < \nu_i \le \ell} \lambda_{\nu_1} \cdot \cdots \cdot \lambda_{\nu_i}$$

is the ith elementary symmetric function of the eigenvalues. The following result generalizes Lemma 16.10 to this context. The Chern classes are characterized by the following properties. The first two properties summarize the discussion above, the next is immediate from the definition, and the final assertion follows from Theorem 19.10.

Lemma 16.11   Let ∇_V be a connection on a complex vector bundle V over a compact manifold M.

1. $dc_k(V, R_\nabla) = 0$ and $[c_k(V, R_\nabla)] \in H^{2k}(M, \mathbb{C})$ is independent of the choice of ∇_V.
2. $c(V \oplus W, R_{\nabla_V \oplus \nabla_W}) = c(V, R_{\nabla_V}) c(W, R_{\nabla_W})$.
3. If $f : N \to M$, then $c(f^*V, R_{f^*\nabla_V}) = f^* c(V, R_{\nabla_V})$.
4. If $\mathbb{L}_{\mathbb{C}}$ is the classifying bundle over $\mathbb{CP}^1$, then $\int_{\mathbb{CP}^1} c_1(\mathbb{L}_{\mathbb{C}}, R_{\nabla_{\mathbb{L}}}) = 1$.
5. Let $P(c_1, c_2, \dots)$ be a polynomial with integer coefficients which is graded homogeneous of degree m. If M has dimension 2m, then

$$\int_M P(c_1(V, R_{\nabla_V}), \dots, c_k(V, R_{\nabla_V})) \in \mathbb{Z}.$$

There are additional characteristic classes which will play an important role which are best described in terms of *generating functions*. If $P(\lambda_1, \dots, \lambda_\ell)$ is a symmetric function of the eigenvalues, then we can express P in terms of the elementary symmetric functions to define an invariant polynomial so $P(A) = P(c_1(A), \dots, c_\ell(A))$. Let $P(\nabla) = P(\frac{\sqrt{-1}}{2\pi} R_\nabla)$ define the characteristic class $P(V) \in H^*_{\mathrm{deR}}(M, \mathbb{C})$.

16.1.6.c The Chern Character. This characteristic class plays a central role in the Hirzebruch–Riemann–Roch Theorem (see Theorem 18.34); it also plays a role in the proof (see Theorem 18.25) that the torus does not admit a metric of positive scalar curvature. It is defined by setting

$$\mathrm{ch}(A) = \sum_{i=1}^{\ell} e^{\lambda_i} = \sum_{i=1}^{\ell} \sum_{k=0}^{\infty} \frac{1}{k!} \lambda_i^k \quad \text{and} \quad \mathrm{ch}(R_\nabla) = \ell + \sum_{k=1}^{\infty} \frac{1}{(2\pi\sqrt{-1})^k k!} \mathrm{Tr}(R_\nabla^k).$$

We expand $\mathrm{ch}(R_\nabla) = \sum_k \mathrm{ch}_k(R_\nabla)$. We have $\mathrm{ch}_k(V) = [\mathrm{ch}_k(R_\nabla)] \in H^{2k}(M; \mathbb{C})$ is independent of the choice of the connection ∇_V. We have, for example,

$$\mathrm{ch}_0 = \ell, \quad \mathrm{ch}_1 = c_1, \quad \mathrm{ch}_2 = \frac{c_1^2 - 2c_2}{2}, \quad \mathrm{ch}_3 = \frac{c_1^3 - 3c_1 c_2 + 3c_3}{6}.$$

Let $K(M)$ be the complex K-theory of M (see Karoubi [41]). The following is well known.

Theorem 16.12

1. $\mathrm{ch}_0(V) = \dim(V)$, $\mathrm{ch}_1(V) = c_1(V)$, and $\mathrm{ch}(f^*V) = f^* \mathrm{ch}(V)$.
2. $\mathrm{ch}(V \oplus W) = \mathrm{ch}(V) + \mathrm{ch}(W)$ and $\mathrm{ch}(V \otimes W) = \mathrm{ch}(V) \mathrm{ch}(W)$.
3. The properties of Assertions 1 and 2 characterize ch.
4. ch defines a natural ring isomorphism from $K(M) \otimes_{\mathbb{R}} \mathbb{C}$ to $\oplus_k H^{2k}(M, \mathbb{C})$.

16.1.6.d The Todd Genus. This characteristic class plays a central role in the Riemann–Roch Theorem (see the discussion in Section 18.4.9). It is defined by

$$\mathrm{Td}(A) = \prod_{i=1}^{\ell} \frac{\lambda_i}{1 - e^{-\lambda_i}} = \prod_{i=1}^{\ell} \left\{ 1 + \frac{\lambda_i}{2} + \frac{\lambda_i^2}{12} - \frac{\lambda_i^4}{720} + \cdots \right\},$$

$$\mathrm{Td}_0 = 1, \quad \mathrm{Td}_1 = \frac{c_1}{2}, \quad \mathrm{Td}_2 = \frac{c_1^2 + c_2}{12}, \quad \mathrm{Td}_3 = \frac{c_1 c_2}{24}.$$

We will discuss complex projective space $\mathbb{CP}^m$ subsequently in Section 19.2.2; $\mathbb{CP}^m$ is a holomorphic manifold and the tangent bundle has a natural complex structure we denote by $T_c(\mathbb{CP}^m)$. The Todd class is characterized by Assertions 1 and 2 in the following result.

Theorem 16.13

1. $\mathrm{Td}_0(V) = 1$ and $\mathrm{Td}(V \oplus W) = \mathrm{Td}(V)\,\mathrm{Td}(W)$.
2. $\int_{\mathbb{CP}^m} \mathrm{Td}_m(T_c(\mathbb{CP}^m)) = 1$.

16.1.6.e The Pontrjagin classes. If V is a real vector bundle, we can complexify V and set $V_{\mathbb{C}} := V \otimes_{\mathbb{R}} \mathbb{C}$. It turns out $c_{2i+1}(V_{\mathbb{C}}) = 0$ and we define the jth *Pontrjagin class* by setting

$$p_j(V) := (-1)^j c_{2j}(V).$$

16.1.6.f The $\hat{A}$ genus. This characteristic class will play an important role in our study of spinors in Section 18.3.6. It is given by the generating function $\frac{\lambda}{e^{\lambda/2} - e^{-\lambda/2}}$. We have, for example,

$$\hat{A}_0 = 1, \quad \hat{A}_1 = -\frac{p_1}{24}, \quad \hat{A}_2 = \frac{-4p_2 + 7p_1^2}{5760}, \quad \hat{A}_3 = \frac{-16p_3 + 44p_2 p_1 - 31p_1^3}{967680}.$$

We refer to Hirzebruch [38] for further details.

16.1.6.g The Euler form. This characteristic class properly speaking is a SO characteristic class and plays an important role in the Chern–Gauss–Bonnet Theorem (see Theorem 18.10). We define the *Euler form* or *Pfaffian* in dimension $m = 2\mathrm{m}$ by

$$E_\mathrm{m} := \frac{(-1)^\mathrm{m}}{8^\mathrm{m} \pi^\mathrm{m} \mathrm{m}!} g(e^{i_1} \wedge \cdots \wedge e^{i_\mathrm{m}}, e^{j_1} \wedge \cdots \wedge e^{j_\mathrm{m}}) R_{i_1 i_2 j_1 j_2} \cdots R_{i_{\mathrm{m}-1} i_\mathrm{m} j_{\mathrm{m}-1} j_\mathrm{m}}.$$

One has, for example, $E_2 := (4\pi)^{-1}\tau$, and $E_4 := (32\pi^2)^{-1}\{\tau^2 - 4\|\rho\|^2 + \|R\|^2\}$ where ρ is the Ricci tensor. Similar formulas for E_6 and E_8 can be found in Pekonen [57].

16.1.7 THE BAIRE CATEGORY THEOREM [7].

Theorem 16.14   Let $C_1 \subset C_2 \subset \cdots$ be a countable collection of closed subsets of a complete metric space X so that $X = \cup_n C_n$. Then the interior of C_n is non-empty for some n.

Proof. Suppose to the contrary that C_n has empty interior for all n; we argue for a contradiction. Since the interior of C_1 is empty, $C_1 \neq X$. Choose $P_1 \notin C_1$. As C_1 is closed, we may choose $0 < \varepsilon_1 < \frac{1}{4}$ so $d(P_1, C_1) \geq \varepsilon_1$. Since P_1 is not in the interior of C_2, we may choose $P_2 \notin C_2$ so $d(P_1, P_2) < \frac{1}{4}\varepsilon_1$. As C_2 is closed, we may choose $0 < \varepsilon_2 < \frac{1}{4}\varepsilon_1$ so $d(P_2, C_2) \geq \varepsilon_2$. Continue in this fashion to construct a sequence of points $P_n \in X$ and real numbers ε_n so $0 < \varepsilon_{n+1} < \frac{1}{4}\varepsilon_n$, $d(P_n, C_n) \geq \varepsilon_n$, and $d(P_n, P_{n+1}) < \frac{1}{4}\varepsilon_n < 4^{-n-1}$. Consequently, $\{P_n\}$ is a Cauchy sequence and we may set P_∞ to be the limiting point since X is complete. We then have

$$d(P_n, P_\infty) \leq \varepsilon_n(\tfrac{1}{4} + \tfrac{1}{16} + \cdots) < \varepsilon_n$$

so $d(P_\infty, C_n) > 0$. Thus, $P_\infty \notin C_n$ for any n which is false since $X = \cup_n C_n$. $\square$

16.1.8 ARZELÀ–ASCOLI THEOREM [3–5].
A sequence $\{f_n\}$ of continuous functions defined on a metric space X is said to be *uniformly bounded* if there exists a constant C so that $|f_n(x)| \leq C$ for all n and all $x \in X$. The family is said to be *equi-continuous* if given $\varepsilon > 0$, there exists a uniform $\delta > 0$ so that if $d(x, y) < \delta$, then $|f_n(x) - f_n(y)| < \varepsilon$ for all n and $x, y \in X$. The following result is called the Arzelà–Ascoli Theorem.

Theorem 16.15   Let $\{f_n\}$ be a uniformly bounded equi-continuous sequence of continuous functions on a compact metric space X. Then there exists a subsequence so that f_{n_k} converges uniformly to a continuous function f_∞ on X.

Proof. Since X is compact, there exists a countable dense subset $\{x_k\}$ in X. As the sequence $\{f_n(x_1)\}$ is bounded, there is a subsequence n_i so $\lim_{i \to \infty} f_{n_i}(x_1) = f_\infty(x_1)$ exists. Since the sequence $\{f_{n_i}(x_2)\}$ is bounded, we can choose a subsequence n_{i_j} so $\lim_{j \to \infty} f_{n_{i_j}}(x_2) = f_\infty(x_2)$ exists. We continue in this fashion. Taking the diagonal subsequence then creates a subsequence, which by changing notation we may assume is the original sequence, so

$$\lim_{n \to \infty} f_n(x_k) = f_\infty(x_k) \quad \text{exists for all } k \,.$$

Let $\varepsilon > 0$ be given. Choose $\delta > 0$ so if $d(x, y) < \delta$ then $|f_n(x) - f_n(y)| < \frac{1}{3}\varepsilon$ for all n. The open balls $B_\delta(x_k)$ of radius δ about the points x_k cover X since $\{x_k\}$ is dense in X. Since X is compact, we can extract a finite cover and choose K so that

$$X = \cup_{k \leq K} B_\delta(x_k) \,.$$

Choose N uniform so $|f_n(x_k) - f_\infty(x_k)| < \frac{1}{3}\varepsilon$ for $n \geq N$ and $k \leq K$. Let $x \in X$. Choose x_k so $d(x, x_k) < \delta$ and $k \leq K$. Then if $n, m > N$ we may estimate:

$$
\begin{aligned}
|f_n(x) - f_m(x)| &\leq |f_n(x) - f_n(x_k)| + |f_n(x_k) - f_m(x_k)| + |f_m(x_k) - f_m(x)| \\
&< \tfrac{1}{3}\varepsilon + \tfrac{1}{3}\varepsilon + \tfrac{1}{3}\varepsilon = \varepsilon \,.
\end{aligned}
$$

Consequently, the sequence $\{f_n\}$ is uniformly Cauchy and hence converges uniformly to a limiting function f_∞ which is continuous.   □

16.2   BANACH AND FRÉCHET SPACES

We introduce normed linear spaces $(B, \|\|)$ and state the Stone–Weierstrass Theorem in Section 16.2.1. The operator norm $\|\|\|$ is defined in Section 16.2.2. A normed linear space $(B, \|\|)$ is said to be a Banach space if it is complete; any normed linear space can be completed in a natural fashion to a Banach space. Let B_1 and B_2 be Banach spaces. Let $\mathrm{Hom}(B_1, B_2)$ be the space of continuous linear maps from B_1 to B_2; we will show $(\mathrm{Hom}(B_1, B_2), \|\|\|)$ is a Banach space. Let $\mathrm{GL}(B_1) \subset \mathrm{Hom}(B_1, B_1)$ be the subset of bijective maps. We establish the Banach–Schauder Theorem or Inverse Boundedness Theorem in Section 16.2.3; if $\Psi \in \mathrm{Hom}(B_1, B_2)$ is bijective, then Ψ^{-1} is continuous. We use this result and the Neumann series to show that $\mathrm{GL}(B_1)$ is a topological group and that $\mathrm{GL}(B_1)$ is an open subset of $\mathrm{Hom}(B_1, B)$. Fréchet spaces are discussed briefly in Section 16.2.4.

16.2.1 NORMED LINEAR SPACES.
Let B be an infinite-dimensional vector space (real or complex) which is equipped with a norm $\|\|$, i.e., a non-negative real-valued function on B so that

(1) $\|v\| = 0$ if and only if $v = 0$.

(2) $\|\lambda v\| = |\lambda| \, \|v\|$ for any scalar λ.

(3) $\|v + w\| \le \|v\| + \|w\|$.

The pair $(B, \|\|)$ is said to be a *normed vector space*. Two norms $\|\|_1$ and $\|\|_2$ on B will be said to be *equivalent norms* if there exist positive constants C_1 and C_2 so that

$$\|v\|_1 \le C_1 \|v\|_2 \quad \text{and} \quad \|v\|_2 \le C_2 \|v\|_1$$

for any $v \in B$. Any two norms on a finite-dimensional vector space are equivalent (see Lemma 1.2 of Book I); this is not, of course, the case for an infinite-dimensional vector space.

Let $d(v, w) = \|v - w\|$ give a normed linear space $(B, \|\|)$ the structure of a metric space; B is said to be a *Banach space* if B is complete with respect to this metric. If $\phi \in B$ and $r > 0$, let

$$D_{B,r}(\phi) := \{\psi \in B : \|\psi - \phi\| < r\}.$$

If B_0 is a normed linear space, we can complete B_0 to define a Banach space B and an isometric embedding of B_0 as a dense subset of B.

Example 16.16   Let $C(X)$ be the space of continuous complex-valued functions on a compact metric space X. If $f \in C(X)$, set $\|f\|_\infty = \sup_{x \in X} |f(x)|$. Since the uniform limit of continuous functions is again a continuous function, $\|\|_\infty$ defines a complete norm on $C(X)$ and $(C(X), \|\|_\infty)$ is a Banach space.

We have the Stone–Weierstrass Theorem [70, 75].

Theorem 16.17   Let $\mathfrak{A}$ be a unital complex subalgebra of $C(X)$ which separates points of X, and which is closed under conjugation. Then $\mathfrak{A}$ is dense in $C(X)$.

16.2.2 THE OPERATOR NORM. Let $(B_1, \|\cdot\|_{B_1})$ and $(B_2, \|\cdot\|_{B_2})$ be normed linear spaces. Let $\Psi : B_1 \to B_2$ be a linear map. We say that Ψ is *bounded* if there exists κ so that

$$\|\Psi\phi\|_{B_2} \leq \kappa \|\phi\|_{B_1} \quad \text{for every} \quad \phi \in B_1 \, .$$

If Ψ is bounded, then the *operator norm* $\|\|\Psi\|\|$ is defined by:

$$\|\|\Psi\|\|_{B_1,B_2} := \sup_{\phi \in B_1, \|\phi\|_{B_1} = 1} \|\Psi\phi\|_{B_2} < \infty \, .$$

We shall often set $\|\|\cdot\|\| = \|\|\cdot\|\|_{B_1,B_2}$ to simplify the notation. It is immediate that

$$\|\|\Psi \circ \Phi\|\| \leq \|\|\Psi\|\| \, \|\|\Phi\|\| \quad \text{and} \quad \|\|a_1\Psi_1 + a_2\Psi_2\|\| \leq |a_1| \, \|\|\Psi_1\|\| + |a_2| \, \|\|\Psi_2\|\| \, . \qquad (16.2.a)$$

Let $\mathrm{Hom}(B_1, B_2)$ be the set of bounded linear maps from B_1 to B_2. By Equation (16.2.a), $(\mathrm{Hom}(B_1, B_2), \|\|\cdot\|\|)$ is a normed linear space and composition is a continuous map from $\mathrm{Hom}(B_2, B_3) \times \mathrm{Hom}(B_1, B_2) \to \mathrm{Hom}(B_1, B_3)$. Let $\mathrm{GL}(B) \subset \mathrm{Hom}(B, B)$ be the set of bijective bounded linear maps from B to itself.

Lemma 16.18   Let $(B_1, \|\cdot\|_{B_1})$ and $(B_2, \|\cdot\|_{B_2})$ be normed linear spaces. Let $\Psi : B_1 \to B_2$ be linear. The following assertions are equivalent.

1. Ψ is uniformly continuous.

2. Ψ is continuous at 0.

3. Ψ is bounded.

Proof. It is clear that Ψ is uniformly continuous implies Ψ is continuous at the origin. Suppose Ψ is continuous at 0. Let $\varepsilon := 1$. We have $\Psi(0) = 0$. Choose $\delta > 0$ so that $\|\phi\|_{B_1} < \delta$ implies $\|\Psi\phi\|_{B_2} < 1$. Rescaling yields $\|\phi\|_{B_1} \leq 1$ implies $\|\Psi\phi\|_{B_2} \leq \delta^{-1}$ and hence Ψ is bounded. Suppose Ψ is bounded. Then $\|\Psi(\phi_1 - \phi_2)\|_{B_2} \leq \|\|\Psi\|\| \, \|\phi_1 - \phi_2\|_{B_1}$ and Ψ is uniformly continuous. □

We will often define Banach spaces as completions of an initial linear space with respect to various norms. Bounded linear operators will initially often only be defined on the initial linear space; Assertion 1 in the following lemma permits us to extend them to the completions; by Lemma 16.18 such maps will be continuous.

Lemma 16.19

1. Let $(B_1, \|\cdot\|_{B_1})$ and $(B_2, \|\cdot\|_{B_2})$ be normed linear spaces and let $\Psi : B_1 \to B_2$ be a continuous linear map. Then Ψ extends to a bounded linear map with the same operator norm from the completion of B_1 to the completion of B_2.

2. If B_1 and B_2 are Banach spaces, then $(\mathrm{Hom}(B_1, B_2), \|\|\cdot\|\|)$ is a Banach space.

Proof. Assertion 1 is immediate from the definition. We argue as follows to prove Assertion 2. Suppose that $\{\Psi_n\}$ is a sequence of bounded linear maps from B_1 to B_2 so that

$$\lim_{n,m \to \infty} \|\|\Psi_n - \Psi_m\|\| = 0 \,.$$

Let $\varepsilon > 0$ be given. Choose $N = N(\varepsilon)$ so

$$n, m \geq N \quad \text{implies} \quad \|\|\Psi_n - \Psi_m\|\| < \varepsilon \,.$$

In particular, there exists $\kappa > 0$ so $\|\|\Psi_n\|\| \leq \kappa$ for all n. Let $\phi \in B_1$. If $n, m > N$, then

$$\|\Psi_n \phi - \Psi_m \phi\|_{B_2} \leq \|\|\Psi_n - \Psi_m\|\| \, \|\phi\|_{B_1} \leq \varepsilon \|\phi\|_{B_1} \,.$$

Thus, $\{\Psi_n \phi\}$ is a Cauchy sequence in B_2. Since B_2 is complete, we may set

$$\Psi_\infty \phi = \lim_{n \to \infty} \Psi_n \phi \,.$$

It is immediate that Ψ_∞ is linear and that $\|\Psi_\infty \phi\|_{B_2} \leq \kappa \|\phi\|_{B_1}$ and hence $\|\|\Psi_\infty\|\| \leq \kappa$; Ψ is a bounded linear map. Finally, a similar computation shows

$$\lim_{n \to \infty} \|\|\Psi_n - \Psi_\infty\|\| = 0$$

and hence $(\mathrm{Hom}(B_1, B_2), \|\|\cdot\|\|)$ is complete. $\square$

16.2.3 THE BANACH–SCHAUDER THEOREM.

Theorem 16.20 Let B_1 and B_2 be Banach spaces. If Ψ is a bounded bijective linear map from B_1 to B_2, then Ψ^{-1} is continuous.

Proof. Let

$$D_{r,B_i}(0) := \{x_i \in B_i : \|x_i\| < r\} \quad \text{and} \quad C_n := \overline{\Psi D_{n,B_1}(0)}$$

for i=1,2, and for $n = 1, 2, \ldots$. Since Ψ is surjective, $B_2 = \cup_n C_n$. By Theorem 16.14, the interior of C_n is non-empty for some n. Choose $\phi \in B_2$ and $\delta > 0$ so $D_{\delta, B_2}(\phi) \subset C_n$. Choose k so $-\phi \in C_k$. Then

$$B_{\delta, B_2}(0) = B_{\delta, B_2}(\phi) - \phi \subset C_n + C_k \subset C_{n+k} \,. \tag{16.2.b}$$

We may rescale Equation (16.2.b) to see that $B_{\delta t, B_2}(0) \subset \overline{\Psi D_{(n+k)t, B_2}(0)}$ for any $t > 0$. Let $s := (n + k)/\delta$. Given $r \in \mathbb{R}^+$, choose t so $r = \delta t$. We then have

$$B_{r, B_2}(0) \subset \overline{\Psi D_{rs, B_1}(0)} \quad \text{for any} \quad r \in \mathbb{R}^+. \tag{16.2.c}$$

Let $\phi_0 \in B_2$ satisfy $\|\phi_0\|_{B_2} \leq 1$. By Equation (16.2.c) with $r = 1$, there exists $\psi_0 \in B_1$ so

$$\|\psi_0\|_{B_1} \leq s \quad \text{and} \quad \|\phi_0 - \Psi\psi_0\|_{B_2} \leq 4^{-1}.$$

Set $\phi_1 = \phi_0 - \Psi\psi_0$. By Equation (16.2.c) with $r = 4^{-1}$, there exists $\psi_1 \in B_1$ so

$$\|\psi_1\|_{B_1} \leq 4^{-1}s \quad \text{and} \quad \|\phi_1 - \Psi\psi_1\|_{B_2} \leq 4^{-2}.$$

Continue to construct a sequence of elements $\phi_n \in B_2$ and $\psi_n \in B_1$ so

$$\phi_n = \phi_{n-1} - \Psi\psi_{n-1} \quad \text{for} \quad n \geq 1,$$
$$\|\psi_n\|_{B_1} \leq 4^{-n}s \quad \text{and} \quad \|\phi_n - \Psi\psi_n\|_{B_2} \leq 4^{-n-1} \quad \text{for} \quad n \geq 0.$$

Since B_1 is a Banach space, we may set $\psi_\infty := \lim_{n\to\infty}\{\psi_0 + \cdots + \psi_n\} \in B_1$. We may express $\phi_0 = \Psi\psi_0 + \cdots + \Psi\psi_n + \phi_{n+1}$. Since $\lim_{n\to\infty} \phi_n = 0$ and since Ψ is continuous, we conclude $\phi_0 = \Psi\psi$. We have $\|\psi\|_{B_1} \leq s(1 + 4^{-1} + 4^{-2} + \cdots) \leq 2s$. Consequently,

$$D_{1, B_2}(0) \subset \Psi D_{2s, B_1}(0). \tag{16.2.d}$$

Since Ψ is bijective, Ψ^{-1} is a well defined linear map. We apply Ψ^{-1} to Equation (16.2.d) to see $\Psi^{-1}D_{1, B_2}(0) \subset D_{2s, B_1}(0)$. It now follows that $\|\|\Psi^{-1}\|\| \leq 2s$. $\qquad\square$

We use Theorem 16.20 to establish the following result.

Lemma 16.21   Let B be a Banach space. Then $\mathrm{GL}(B)$ is an open subset of $\mathrm{Hom}(B)$ and $(\mathrm{GL}(B), \|\|)$ is a topological group.

Proof. Note that $\mathrm{Hom}(B, B)$ is a Banach space by Lemma 16.19. By definition, $\mathrm{GL}(B)$ is the set of continuous bijective linear transformations of B. Thus, if Ψ belongs to $\mathrm{GL}(B)$, then Ψ^{-1} is continuous by Theorem 16.20 and $\Psi^{-1} \in \mathrm{GL}(B)$. Consequently, composition makes $\mathrm{GL}(B)$ into a group. We apply Equation (16.2.a). The composition of bounded linear maps is again a bounded linear map and composition gives a continuous map from $\mathrm{GL}(B) \times \mathrm{GL}(B)$ to $\mathrm{GL}(B)$. Let $\mathrm{Inv}(\Phi) = \Phi^{-1}$. Since $\mathrm{Inv}(\Psi \circ \Phi) = \mathrm{Inv}(\Phi) \circ \mathrm{Inv}(\Psi)$, to show that Inv is continuous on all of $\mathrm{GL}(B)$, it suffices to prove Inv is continuous at Id. Let $L_\Phi\Psi = \Phi \circ \Psi$ for $\Phi \in \mathrm{GL}(B)$. Since $L_\Phi^{-1} = L_{\Phi^{-1}}$, L_Φ is a homeomorphism of $\mathrm{Hom}(B, B)$. Thus, to show $\mathrm{GL}(B)$ is an open subset of $\mathrm{Hom}(B, B)$, it suffices to show $\mathrm{GL}(B)$ is an open neighborhood of Id.

We compute at Id. Suppose $\Psi \in B_{\frac{1}{2}}(\mathrm{Id})$. Let $C(\Psi) := \mathrm{Id} - \Psi$. We define

$$\Theta(\Psi) := \mathrm{Id} + C(\Psi) + \cdots + C(\Psi)^n + \cdots.$$

This is called the *Neumann series*. Since $\|\|C(\Psi)\|\| \leq \frac{1}{2}$, $\|\|C(\Psi)^n\|\| \leq \|\|C(\Psi)\|\|^n < 2^{-n}$ by Equation (16.2.a) so $\Theta(\Psi)$ is well defined. The estimate $\|\|C(\Psi)^n\|\| \leq 2^{-n}$ justifies the following computation:

$$\Theta(\Psi) \cdot \Psi = \sum_{n=0}^{\infty} C(\Psi)^n (\mathrm{Id} - C(\Psi)) = \sum_{n=0}^{\infty} C(\Psi)^n - \sum_{n=1}^{\infty} C(\Psi)^n = \mathrm{Id},$$

$$\Psi \cdot \Theta(\Psi) = (\mathrm{Id} - C(\Psi)) \sum_{n=0}^{\infty} C(\Psi)^n = \sum_{n=0}^{\infty} C(\Psi)^n - \sum_{n=1}^{\infty} C(\Psi)^n = \mathrm{Id} .$$

This shows Ψ is invertible and that $\Theta(\Psi) = \mathrm{Inv}(\Psi)$. Consequently, $B_{\frac{1}{2}}(\mathrm{Id}) \subset \mathrm{GL}(B)$ and $\mathrm{GL}(B)$ is a neighborhood of Id. Furthermore,

$$||| \mathrm{Id} - \Theta(\Psi) ||| \leq \sum_{n=1}^{\infty} ||| \mathrm{Id} - \Psi |||^n = \frac{||| \mathrm{Id} - \Psi |||}{1 - ||| \mathrm{Id} - \Psi |||} \leq 2 ||| \mathrm{Id} - \Psi ||| .$$

Consequently, the map $\Psi \to \Theta(\Psi)$ is continuous at the identity. $\square$

16.2.4 FRÉCHET SPACES. We refer to Conway [18] or Treves [72] for further details. We begin with a useful observation.

Lemma 16.22   Let (X, d) be a metric space. Set $\tilde{d}(x, y) = \dfrac{d(x, y)}{1 + d(x, y)}$. Then $(X, \tilde{d})$ is a metric space with the same topology.

Proof. It is clear that $\tilde{d}(x, y) = \tilde{d}(y, x)$ and that $\tilde{d}(x, y) \geq 0$ with equality if and only if $x = y$. Thus, to show $\tilde{d}$ is a distance function, we must verify the triangle inequality. Let $f(a) := \frac{a}{1+a}$. As $\frac{f(a)}{a} = \frac{1}{1+a}$ is monotonically decreasing, $\frac{f(a+b)}{a+b} \leq \frac{f(a)}{a}$ so

$$f(a + b) = \frac{a f(a + b)}{a + b} + \frac{b f(a + b)}{a + b} \leq a \frac{f(a)}{a} + b \frac{f(b)}{b} = f(a) + f(b) . \qquad (16.2.e)$$

Since $f(a) = 1 - \frac{1}{1+a}$ is monotone increasing, $a < b$ implies $f(a) < f(b)$. Consequently, the triangle inequality for d implies

$$\tilde{d}(x, y) = f(d(x, y)) \leq f(d(x, z) + d(z, y)) . \qquad (16.2.f)$$

We now use Equations (16.2.e) and (16.2.f) to see

$$\tilde{d}(x, y) \leq f(d(x, z) + d(z, y)) \leq f(d(x, z)) + f(d(z, y)) = \tilde{d}(x, z) + \tilde{d}(z, y) .$$

This establishes the triangle inequality for $\tilde{d}$. We have $\tilde{d}(x, y) < d(x, y)$. We express $d = \frac{\tilde{d}}{1 - \tilde{d}}$ to see that if $\tilde{d}(x, y) < \frac{1}{2}$, then $d(x, y) < 2\tilde{d}(x, y)$. The requisite equivalence of topologies now follows. $\square$

Definition 16.23   Let $\mathfrak{F} := (F, \|\cdot\|_1, \|\cdot\|_2, \dots)$ where F is a complex vector space and where $\{\|\cdot\|_k\}_{k=1,2,\dots}$ is a countable sequence of norms on F with $\|\cdot\|_1 \leq \|\cdot\|_2 \leq \dots$. We give F a topology $\mathfrak{T}$ by requiring that a subset U of F is open if and only if for every point $u \in U$, there exists a positive integer $k = k(u)$ and there exists $\varepsilon > 0$ so that the set $\{v \in F : \|v - u\|_k < \varepsilon\}$ is contained in U; the topology $\mathfrak{T}$ is characterized by the fact that the functions $u \to \|u\|_k$ are continuous maps from F to $\mathbb{R}$ for $k = 1, 2, \dots$. In other words, if $\{f_n\}$ is a sequence of elements in F, then f_n converges to f in $\mathfrak{T}$ if and only if f_n converges to f in each of the norms $\|\cdot\|_k$. We shall show in Lemma 16.24 that $\mathfrak{T}$ is metrizable. We say that $\mathfrak{F}$ is a *Fréchet space* if $\mathfrak{T}$ is complete.

Lemma 16.24

1. Any Fréchet space is metrizable.
2. A linear map from a Fréchet space $\mathfrak{F}_1$ to a Fréchet space $\mathfrak{F}_2$ is continuous if and only if for every n, there exists $N(n)$ so $\|\cdot\|_{k,F_2} \leq \|\cdot\|_{k,F_1}$.

Proof. Let $\mathfrak{F}$ be a Fréchet space. Set $d(x, y) = \sum_{k=1}^{\infty} 2^{-k} \dfrac{\|x - y\|_k}{1 + \|x - y\|_k}$. Assertion 1 follows from Lemma 16.22; Assertion 2 is immediate.   □

Example 16.25   If $M = [0, 1]$ and $\phi \in C^k[0, 1]$, set $\|\phi\|_k := \sum_{j=0}^{k} \sup_{x \in [0,1]} |\phi^{(j)}(x)|$. Then $(C^k[0, 1], \|\cdot\|_k)$ is complete so this gives $C^k[0, 1]$ the structure of a Banach space. However, $C^\infty[0, 1]$ is not a Banach space; the topology is defined by the countable collection of increasing norms $\|\cdot\|_k$ so $C^\infty[0, 1]$ is a Fréchet space. We will generalize this example subsequently in Section 17.3.1.

16.3   HILBERT SPACES

In Section 16.3.1, we define inner product spaces and prove the Cauchy–Schwarz–Bunyakovsky inequality. In Section 16.3.2, we derive Peetre's inequality. We will use these inequalities subsequently in Section 17.2 in our discussion of Sobolev spaces. We present the Gram–Schmidt process in Section 16.3.3. We establish Parseval's inequality in Section 16.3.4, the Hilbert Projection Theorem in Section 16.3.5, and the Riesz–Fréchet Representation Theorem in Section 16.3.6.

16.3.1 THE CAUCHY–SCHWARZ–BUNYAKOVSKY INEQUALITY.   An *inner product space* is a pair $(H, (\cdot, \cdot))$ where H is a complex vector space and $(\cdot, \cdot)$ is a positive definite Hermitian inner product on H. Set $\|\phi\| = \sqrt{(\phi, \phi)}$. In the following lemma, Assertion 1 is the *Cauchy–Schwarz–Bunyakovsky inequality* [12, 15, 66], Assertion 2 is the *triangle inequality*, and Assertion 3 is *interpolation*.

Lemma 16.26   Let ϕ and ψ belong to an inner product space $(H, (\cdot, \cdot))$.

1. $|(\phi, \psi)| \leq \|\phi\| \, \|\psi\|$.
2. $\|\phi + \psi\| \leq \|\phi\| + \|\psi\|$.
3. $|(\phi, \psi)| \leq \varepsilon^2 \|\phi\|^2 + \varepsilon^{-2} \|\psi\|^2$ for any $\varepsilon > 0$.
4. $(H, \|\|)$ is a normed linear space.

Proof. By replacing ψ by $e^{\sqrt{-1}\theta} \psi$ for some suitably chosen θ, we may assume without loss of generality that (ϕ, ψ) is real in proving Assertion 1. Thus, $(\psi, \phi) = (\phi, \psi)$. Let

$$f(t) := \|t\phi + \psi\|^2 = t^2 \|\phi\|^2 + 2t(\phi, \psi) + \|\psi\|^2 \,.$$

Since $f(t)$ is a non-negative quadratic function of t, the discriminant is non-positive so

$$4(\phi, \psi)^2 - 4\|\psi\|^2 \|\phi\|^2 \leq 0 \,;$$

Assertion 1 follows. We use Assertion 1 to establish Assertion 2 by computing:

$$\begin{aligned}
\|\phi + \psi\|^2 &= (\phi + \psi, \phi + \psi) = (\phi, \phi) + (\phi, \psi) + (\psi, \phi) + (\psi, \psi) \\
&\leq \|\phi\|^2 + 2\|\phi\| \, \|\psi\| + \|\psi\|^2 = (\|\phi\| + \|\psi\|)^2 \,.
\end{aligned}$$

As $(|a| - |b|)^2 \geq 0$, $|a|^2 + |b|^2 - 2|a| \, |b| \geq 0$ so

$$|a| \, |b| \leq 2|a| \, |b| \leq |a|^2 + |b|^2 \,. \tag{16.3.a}$$

By Assertion 1 and Equation (16.3.a), $|(\phi, \psi)| \leq \varepsilon \|\phi\| \, \varepsilon^{-1} \|\psi\| \leq \varepsilon^2 \|\phi\|^2 + \varepsilon^{-2} \|\psi\|^2$. This establishes Assertion 3; Assertion 4 follows from Assertion 2. $\square$

16.3.2 PEETRE'S INEQUALITY.

Lemma 16.27   Let $(H, (\cdot, \cdot))$ be an inner product space. If $\phi, \psi \in H$ and if $s \in \mathbb{R}$, then

1. $(1 + \|\phi\|^2)^s \leq 2^{|s|}(1 + \|\psi\|^2)^s (1 + \|\phi - \psi\|^2)^{|s|}$.
2. $(1 + \|\phi - \psi\|)^{-k} \leq (1 + \|\phi\|)^{-k}(1 + \|\psi\|)^k$ for $k \geq 0$.

Proof. Let $u, v \in H$. By Equation (16.3.a), $2\|u\| \, \|v\| \leq \|u\|^2 + \|v\|^2$. Thus, we may use the Cauchy–Schwarz–Bunyakovsky inequality to estimate

$$\begin{aligned}
1 + \|u - v\|^2 &= 1 + \|u\|^2 + \|v\|^2 - (u, v) - (v, u) \leq 1 + \|u\|^2 + \|v\|^2 + 2\|u\| \, \|v\| \\
&\leq 1 + 2\|u\|^2 + 2\|v\|^2 \leq 2(1 + \|u\|)(1 + \|v\|) \,.
\end{aligned}$$

Set $\phi = u - v$ and $\psi = -v$ to obtain the estimate $1 + \|\phi\|^2 \leq 2(1 + \|\phi - \psi\|)(1 + \|\psi\|)$. Consequently, $(1 + \|\phi\|^2)^{|s|} \leq 2^{|s|}(1 + \|\psi\|)^{|s|}(1 + \|\phi - \psi\|)^{|s|}$. Assertion 1 now follows if $s \geq 0$. If $s \leq 0$, interchange the roles of ϕ and ψ to see

$$(1 + \|\psi\|^2)^{-s} \leq 2^{|s|}(1 + \|\phi\|)^{-s}(1 + \|\phi - \psi\|)^{|s|} \,.$$

Cross multiply to complete the proof of Assertion 1. We have that

$$1 + \|\phi\| \leq 1 + \|\phi - \psi\| + \|\psi\| \leq (1 + \|\phi - \psi\|)(1 + \|\psi\|) \quad \text{so}$$
$$(1 + \|\phi - \psi\|)^{-k} \leq (1 + \|\phi\|)^{-k}(1 + \|\psi\|)^{k} \quad \text{for} \quad k \geq 0. \qquad \square$$

16.3.3 THE GRAM–SCHMIDT PROCESS.

Definition 16.28   Let $\mathcal{H} := (H, (\cdot, \cdot))$ be an inner product space.

1. $\mathcal{H}$ is said to be a *Hilbert space* if $(H, \|\|)$ is complete.

2. $\mathcal{H}$ is said to be *separable* if there is a countable dense subset of H.

3. Let $\delta_{ij} = \left\{ \begin{array}{ll} 1 & \text{if } i = j \\ 0 & \text{if } i \neq j \end{array} \right\}$ be the *Kronecker symbol*.

4. $\{\psi_1, \psi_2, \ldots\}$ is said to be an *orthonormal sequence* if $(\psi_i, \psi_j) = \delta_{ij}$.

5. If $\{\psi_i\}$ is an orthonormal sequence, $\sigma_i(\phi) := (\phi, \psi_i)$ is the ith *Fourier coefficient*.

6. An orthonormal sequence $\{\psi_1, \psi_2, \ldots\}$ is said to be a *complete orthonormal basis* if the closure of the linear span of $\{\psi_n\}$ is all of H.

Example 16.29   Let ℓ^2 be the vector space of all complex sequences $\vec{a} = (a_1, a_2, \ldots)$ so that $\sum_n |a_n|^2 < \infty$. Set $(\vec{a}, \vec{b})_{\ell^2} = \sum_n a_n \bar{b}_n$; this converges by the Cauchy–Schwarz–Bunyakovsky inequality. The associated norm $\|\vec{a}\|^2 = \sum_n |a_n|^2$ is complete so ℓ^2 is a Hilbert space. Let $e_n = (0, \ldots, 0, 1, 0, \ldots)$; $\{e_n\}$ is a complete orthonormal basis for ℓ^2. If S is the set of all finite sequences $(a_1, a_2, \ldots, a_n, 0, \ldots)$ where the a_i are rational, then S is a countable dense subset of ℓ^2. Consequently, ℓ^2 is separable.

The following is called the *Gram–Schmidt process*; see Gram [30].

Construction 16.30   Let $\{\phi_1, \ldots\}$ be linearly independent elements of H. Set

$$\psi_1 := \frac{\phi_1}{\|\phi_1\|} \quad \text{and} \quad \psi_n := \frac{\phi_n - (\phi_n, \phi_1)\phi_1 - \cdots - (\phi_n, \phi_{n-1})\phi_{n-1}}{\|\phi_n - (\phi_n, \phi_1)\phi_1 - \cdots - (\phi_n, \phi_{n-1})\phi_{n-1}\|} \quad \text{for} \quad n \geq 2.$$

One then may check that $\{\psi_1, \ldots\}$ is an *orthonormal sequence*.

16.3.4 PARSEVAL'S INEQUALITY.

Assertion 1 in the following result is called *Parseval's inequality* and Assertion 2 is called *Parseval's identity*. Let ℓ^2 be the infinite-dimensional separable Hilbert space defined in Example 16.29.

Theorem 16.31   Let $(H, (\cdot, \cdot))$ be an infinite-dimensional inner product space.

1. If $\{\psi_i\}_{i=1}^{\infty}$ is an orthonormal subset of H, then $\displaystyle\sum_{i=1}^{\infty} |\sigma_i(\phi)| \leq \|\phi\|^2$ for any $\phi \in H$.

2. If H is a Hilbert space and if $\{\psi_i\}_{i=1}^{\infty}$ is a complete orthonormal basis for H,

$$\phi = \sum_{i=1}^{\infty} \sigma_i(\phi)\psi_i \quad \text{and} \quad (\phi, \psi) = \sum_{i=1}^{\infty} \sigma_i(\phi)\bar{\sigma}_i(\psi) \quad \text{for any} \quad \phi, \psi \in H .$$

3. Any infinite-dimensional separable Hilbert space is isometric to ℓ^2.

Proof. Let $S_n(\phi) = \sigma_1(\phi)\psi_1 + \cdots + \sigma_n(\phi)\psi_n$ be the nth partial sum. If $1 \le \ell \le n$, then

$$(S_n(\phi), S_n(\phi)) = \sum_{i,j=1}^{n} \sigma_i(\phi)\bar{\sigma}_j(\phi)(\psi_i, \psi_j) = \sum_{i=1}^{n} |\sigma_i(\phi)|^2,$$

$$(\phi - S_n(\phi), \psi_\ell) = \sigma_\ell(\phi) - \sum_{i=1}^{n} \sigma_i(\phi)(\psi_i, \psi_\ell) = \sigma_\ell(\phi) - \sigma_\ell(\phi) = 0 .$$

Since $\phi - S_n(\phi)$ is perpendicular to ψ_ℓ for $1 \le i \le \ell$, $\phi - S_n(\phi)$ is perpendicular to $S_n(\phi)$. We establish Assertion 1 by computing

$$\|\phi\|^2 = \|\phi - S_n(\phi)\|^2 + \|S_n(\phi)\|^2 = \|\phi - S_n(\phi)\|^2 + \sum_{i=1}^{n} |\sigma_i(\phi)|^2,$$

$$\sum_{i=1}^{n} |\sigma_i(\phi)|^2 = \|\phi\|^2 - \|\phi - S_n(\phi)\|^2 \le \|\phi\|^2 .$$

Let $\{\psi_i\}$ be a complete orthonormal basis for a Hilbert space H. By Assertion 1, $\sum_i |\sigma_i(\phi)|^2$ is a convergent series. Consequently,

$$\lim_{n,m\to\infty} \|S_n(\phi) - S_m(\phi)\|^2 = \lim_{n,m\to\infty} \sum_{i=\min(n,m)+1}^{\max(n,m)} |\sigma_i(\phi)|^2 = 0 \quad \text{so}$$

$$\phi_\infty := \sum_{i=1}^{\infty} \sigma_i(\phi)\psi_i = \lim_{n\to\infty} S_n(\phi) \quad \text{is well defined.}$$

Since $(\phi - \phi_\infty, \psi_i) = 0$ for all i, $\phi - \phi_\infty$ is perpendicular to the closure of the linear span of the $\{\psi_i\}$ which is, by assumption, all of H. Consequently, $\phi - \phi_\infty = 0$ and Assertion 2 follows.

If H is separable, let S be a countable dense subset of H. After deleting linearly dependent elements of S, the Gram–Schmidt process (see Construction 16.30) can then be applied to construct a complete orthonormal basis $\{\psi_i\}$ for H. By Assertion 2, the map

$$\phi \to (\sigma_1(\phi), \dots, \sigma_n(\phi), \dots)$$

is an isometry from H to ℓ^2.   □

   We can complete any inner product space H_0 to obtain a Hilbert space H so that H_0 is a dense subset of H.

Example 16.32 Let (M, g) be a compact Riemannian manifold. Set

$$(\phi, \psi)_{L^2} = \int_M \phi(x)\bar{\psi}(x)|\operatorname{dvol}| \quad \text{for} \quad \phi, \psi \in C(M). \tag{16.3.b}$$

Let $L^2(M)$ be the completion of $C(M)$ with respect to the norm $\|\cdot\|_{L^2}$; we suppress g from the notation in the interests of notational simplification as the underlying topology is independent of g. By the Whitney Embedding Theorem [77] (see Theorem 2.2 of Book I), M embeds in $\mathbb{R}^k$ for some k. Let $\mathfrak{A} = \mathbb{C}[x^1, \ldots, x^k]$ where $(x^1, \ldots, x^k)$ are the usual coordinates on $\mathbb{R}^k$ and let $\mathfrak{B}$ be the algebra of functions obtained by restricting elements of $\mathfrak{A}$ to M. Then $\mathfrak{B}$ is a unital complex subalgebra of $C(M)$ which separates points of M, and which is closed under complex conjugation. We use Theorem 16.17 to see that $\mathfrak{B}$ is dense in the Banach space $(C(M), \|\cdot\|_\infty)$ of Example 16.16. Since $\|\phi\|_{L^2} \leq \operatorname{Vol}(M) \cdot \|\phi\|_\infty$ for $\phi \in C(M)$, this implies that $\mathfrak{B}$ is dense in $L^2(M)$ as well. Let S be the subalgebra of $\mathfrak{B}$ obtained by restricting polynomials with rational coefficients to M. Then S is a countable dense subset of $L^2(M)$. This shows that $L^2(M)$ is separable and hence $L^2(M)$ is isometric to ℓ^2 by Theorem 16.31. If M is not assumed compact in Example 16.32, then it is necessary to impose decay conditions at infinity to ensure the integral of Equation (16.3.b) converges absolutely; we refer to Section 17.1.1 when $M = \mathbb{R}^m$ and $|\operatorname{dvol}|$ is suitably normalized Lebesgue measure.

Example 16.33 Let S^1 be the unit circle. If $n \in \mathbb{Z}$, set $\psi_n(\theta) := \frac{1}{\sqrt{2\pi}} e^{\sqrt{-1}n\theta} \in C(S^1)$. Then the set $S := \{\psi_0, \psi_{\pm 1}, \psi_{\pm 2}, \ldots\}$ is an orthonormal subset of $L^2(S^1)$. The space $\mathfrak{A}$ of trigonometric polynomials is the linear span of S. Since $\psi_i \psi_j = \frac{1}{\sqrt{2\pi}} \psi_{i+j}$ and $\bar{\psi}_i = \psi_{-i}$, $\mathfrak{A}$ is a unital complex subalgebra of $C(S^1)$ which separates points of S^1 and which is closed under complex conjugation. By Theorem 16.17, $\mathfrak{A}$ is dense in $L^2(S^1)$. Thus, S is a complete orthonormal basis for $L^2(S^1)$. The expansion $\phi = \sum_{n \in \mathbb{Z}} \sigma_n(\phi)\psi_n$ in L^2 is called a *Fourier series*. Note that there exists a continuous function ϕ so this series does not converge in $(C(S^1), \|\cdot\|_0)$. Let $D = -\partial_\theta^2$. Then $D\psi_n = n^2\psi_n$ so S diagonalizes D. If ϕ is smooth, then $\sigma_n(D^\ell\phi) = n^{2\ell}\sigma_n(\phi)$ and hence $\lim_{n\to\infty} n^{2\ell}\sigma_n(\phi) = 0$ for any ℓ; the Fourier coefficients are rapidly decreasing. This shows that $\sum_n \sigma_n(\phi)\psi_n$ converges in the C^k topology for any k; since it converges in L^2 to ϕ, the limit function is ϕ; any smooth function can be approximated in the C^k topology by its Fourier series. More generally, in Theorem 17.22, we will show that if V is any Hermitian vector bundle over a compact Riemannian manifold (M, g) and if D is a self-adjoint operator of Laplace type on $C^\infty(V)$, then there exists a complete orthonormal basis for $L^2(V)$ consisting of smooth eigenfunctions of D. If ϕ is a smooth section, then the Fourier coefficients are rapidly decreasing and the Fourier series converges in the C^∞ topology to ϕ. Fourier series have applications in many contexts; we illustrate this in Remark 19.5.

16.3.5 THE HILBERT PROJECTION THEOREM. A subset C of a linear space is said to be *convex* if given any two elements ϕ_1 and ϕ_2 of C, the line segment joining them,

$$t\phi_1 + (1-t)\phi_2 \quad \text{for} \quad t \in [0,1]$$

is contained in C.

Lemma 16.34   Fix an element ψ of a Hilbert space H.

1. Let C be a closed non-empty convex subset of H. There exists a unique $\phi \in C$ so $\|\phi - \psi\|$ is minimal; ϕ is the point of C closest to ψ.
2. Let C be a closed non-empty linear subspace of H.
   (a) $\phi \in C$ satisfies $\|\phi - \psi\|$ is minimal if and only if $\phi - \psi \perp C$.
   (b) Let $C^\perp := \{\psi_1 \in H : (\phi, \psi_1) = 0 \ \forall \ \phi \in C\}$. Then $C^\perp$ is a closed linear subspace of H and $H = C \oplus C^\perp$.

Proof. Set $\varepsilon := \inf_{\phi \in C} \|\phi - \psi\|$. If η and ξ are any two points of H, then

$$\|\eta - \xi\|^2 = 2\|\eta - \psi\|^2 + 2\|\xi - \psi\|^2 - \|\eta + \xi - 2\psi\|^2 . \tag{16.3.c}$$

Let $\phi_n \in C$ be a minimizing sequence so $\lim_{n \to \infty} \|\phi_n - \psi\| = \varepsilon$. Since C is a convex set, $\frac{1}{2}(\phi_i + \phi_j)$ belongs to C. Consequently,

$$\|\phi_i + \phi_j - 2\psi\|^2 = 4\|\tfrac{1}{2}(\phi_i + \phi_j) - \psi\|^2 \geq 4\varepsilon^2 . \tag{16.3.d}$$

We use Equations (16.3.c) and (16.3.d) with $\eta = \phi_i$ and $\xi = \phi_j$ to see

$$\lim_{i,j \to \infty} \|\phi_i - \phi_j\|^2 \leq \lim_{i,j \to \infty} \{2\|\phi_i - \psi\|^2 + 2\|\phi_j - \psi\|^2 - 4\varepsilon^2\} = 0 .$$

Consequently, $\{\phi_i\}$ is a Cauchy sequence. As $(H, \|)$ is complete, $\phi := \lim_{n \to \infty} \phi_n$ is well defined and satisfies $\|\phi - \psi\| = \varepsilon$. Suppose $\|\tilde{\phi} - \psi\| = \varepsilon$ is another minimizer. We set $\eta = \phi$ and $\xi = \tilde{\phi}$ in Equations (16.3.c) and (16.3.d) to complete the proof of Assertion 1 by estimating

$$0 \leq \|\phi - \tilde{\phi}\|^2 = 2\varepsilon^2 + 2\varepsilon^2 - \|\phi - \tilde{\phi} - 2\psi\|^2 \leq 2\varepsilon^2 + 2\varepsilon^2 - 4\varepsilon^2 = 0 .$$

Let C be a closed linear subspace of H. Choose ϕ as in Assertion 1 so $\|\phi - \psi\|$ is minimal. Let $\tilde{\phi} \in C$. Then

$$f(t) := \|(\phi + t\tilde{\phi}) - \psi\|^2 - \|\phi - \psi\|^2 = t\{(\phi - \psi, \tilde{\phi}) + (\tilde{\phi}, \phi - \psi)\} + t^2\|\tilde{\phi}\|^2 \geq 0 .$$

Replace $\tilde{\phi}$ by $e^{\sqrt{-1}\theta}\tilde{\phi}$ where the angle θ is chosen so that $\left(e^{\sqrt{-1}\theta}\tilde{\phi}, \phi - \psi\right)$ is real. Since $f(t)$ has a minimum at $t = 0$, $(\tilde{\phi}, \phi - \psi) = 0$ so $(\phi - \psi) \perp \tilde{\phi}$ and hence $(\phi - \psi) \perp C$. Conversely, suppose that $(\phi - \psi) \perp C$. Then $\|(\phi + t\tilde{\phi}) - \psi\|^2 - \|\phi - \psi\|^2 = t^2\|\tilde{\phi}\|^2$ and hence ϕ is the minimizer. This proves Assertion 2a. It is clear that $C \cap C^\perp = \{0\}$ and that $C^\perp$ is a closed linear subspace of H. Let $\psi \in H$. Let $\phi \in C$ be the minimizer of Assertion 1. By Assertion 2a, $(\phi - \psi) \in C^\perp$. Since $\psi = \phi + (\psi - \phi)$, $C \oplus C^\perp = H$. $\square$

Construction 16.35   Let C be a closed linear subspace of a separable Hilbert space H. Use the Gram–Schmidt process (see Construction 16.30) to construct an orthonormal basis $\{\phi_1, \dots\}$ for C. We use the decomposition $H = C \oplus C^\perp$ to define the orthogonal projection π_C of H on C and the complementary projection $\pi_C^\perp = \mathrm{Id} - \pi_C$ on $C^\perp$. We then have the *projection formula*

$$\pi_C \phi = \sum_n (\phi, \phi_n)\phi_n \quad \text{and} \quad \pi_C^\perp \phi = \phi - \sum_n (\phi, \phi_n)\phi_n .$$

It is immediate that $\||\pi_C\|| = 1$.

16.3.6 THE RIESZ–FRÉCHET REPRESENTATION THEOREM [22, 64]. Let ψ belong to H. Since $|(\phi, \psi)| \leq \|\phi\| \, \|\psi\|$, the map $L_\psi : \phi \to (\phi, \psi)$ is a bounded linear map from H to $\mathbb{C}$. We now show, conversely, that every bounded linear map from a Hilbert space to $\mathbb{C}$ arises in this fashion.

Theorem 16.36   Let L be a continuous linear map from a Hilbert space H to $\mathbb{C}$. Then there exists a unique $\psi \in H$ so that $L(\phi) = (\phi, \psi)$ for all $\phi \in H$.

Proof. We first establish that ψ is unique. If $L(\phi) = (\phi, \psi) = (\phi, \tilde{\psi})$ for all ϕ, then we have that $(\phi, \psi - \tilde{\psi}) = 0$ for all ϕ and hence $\|\psi - \tilde{\psi}\|^2 = (\psi - \tilde{\psi}, \psi - \tilde{\psi}) = 0$ so $\psi = \tilde{\psi}$. If L is the zero map, then we may take $\psi = 0$. We therefore suppose L is non-trivial and let $C = \ker\{L\} \neq H$. Since L is continuous, C is a closed linear subspace of H. Use Lemma 16.34 to decompose $H = C \oplus C^\perp$ where $C^\perp$ is non-empty. Choose $\psi_1 \in C^\perp$ with $\|\psi_1\| = 1$. Let $\varepsilon := L(\psi_1) \neq 0$. Decompose

$$\phi = \varepsilon^{-1} L(\phi)\psi_1 + (\phi - \varepsilon^{-1} L(\phi)\psi_1) .$$

Since $L(\phi - \varepsilon^{-1} L(\phi)\psi_1) = 0$, $\phi - \varepsilon^{-1} L(\phi)\psi_1 \in C$. Thus, $(\phi - \varepsilon^{-1} L(\phi)\psi_1, \psi_1) = 0$ so

$$(\phi, \psi_1) = (\varepsilon^{-1} L(\phi)\psi_1, \psi_1) + (\phi - L(\phi)\psi_1, \psi_1) = \varepsilon^{-1} L(\phi) + 0 .$$

If we set $\psi := \varepsilon\psi_1$, then $(\phi, \psi) = L(\phi)$ as desired.   □

16.4   SPECTRAL THEORY IN A SEPARABLE HILBERT SPACE

We treat self-adjoint operators in Section 16.4.1 and compact operators in Section 16.4.2. We discuss the spectral theory of compact self-adjoint operators on Hilbert space in Section 16.4.3 and introduce the index in the context of Fredholm operators in Section 16.4.4.

16.4.1 SELF-ADJOINT OPERATORS ON A HILBERT SPACE. Let H be a Hilbert space. Let $\mathrm{Hom}(H) := \mathrm{Hom}(H, H)$ be the space of bounded linear maps of H to itself. Let $\Psi \in \mathrm{Hom}(H)$. Let $E(\lambda) = \{\phi \in H : \Psi\phi = \lambda\phi\}$. If $E(\lambda)$ is non-trivial, then λ is said to be an *eigenvalue* of Ψ and $E(\lambda)$ is said to be the corresponding *eigenspace* of Ψ. An element $0 \neq \phi$ in $E(\lambda)$ is said to be an *eigenvector* of Ψ. We use Theorem 16.36 to define Ψ^* by the identity $(\Psi\phi_1, \phi_2) = (\phi_1, \Psi^*\phi_2)$. The following result shows that the map $\Psi \to \Psi^*$ is a conjugate linear anti-algebra isometry of $\mathrm{Hom}(H)$.

Lemma 16.37   Let H be a Hilbert space. If Ψ and Φ belong to $\mathrm{Hom}(H)$, then

$$(\Psi + \Psi)^* = \Phi^* + \Psi^*, \quad (\Phi \circ \Psi)^* = \Psi^* \circ \Phi^*, \quad (\lambda\Psi)^* = \bar{\lambda}\Psi^*, \quad |||\Psi^*||| = |||\Psi|||.$$

Proof. The first three identities are immediate; we complete the proof by expressing

$$|||\Psi||| = \sup_{\|\phi_1\|=1, \|\phi_2\|=1} |(\Psi\phi_1, \phi_2)| = |||\Psi^*|||.$$ □

Since $|||\Psi^*||| = |||\Psi||| < \infty$, $\Psi^* \in \mathrm{Hom}(H)$. We say that Ψ is *self-adjoint* if $\Psi^* = \Psi$.

Lemma 16.38   Let H be a Hilbert space and let $\Psi \in \mathrm{Hom}(H)$ be self-adjoint.

1. If λ is an eigenvalue of Ψ, then λ is real.
2. If $\lambda \neq \mu$, then $E(\lambda) \perp E(\mu)$.

Proof.   Let λ be an eigenvalue of Ψ. Choose $\phi_\lambda \in E(\lambda)$ with $\|\phi_\lambda\| = 1$. Since Ψ is self-adjoint, we have that $\lambda = (\Psi\phi_\lambda, \phi_\lambda) = (\phi_\lambda, \Psi\phi_\lambda) = \bar{\lambda}$ so $\lambda \in \mathbb{R}$. Let $\phi_\mu \in E(\mu)$ for $\lambda \neq \mu$. Then $\lambda(\phi_\lambda, \phi_\mu) = (\Psi\phi_\lambda, \phi_\mu) = (\phi_\lambda, \Psi\phi_\mu) = \mu(\phi_\lambda, \phi_\mu)$ so $(\lambda - \mu)(\phi_\lambda, \phi_\mu) = 0$. Consequently, $(\phi_\lambda, \phi_\mu) = 0$.   □

16.4.2 COMPACT OPERATORS ON A HILBERT SPACE.

Let H be a Hilbert space and let $\Psi \in \mathrm{Hom}(H)$. Ψ is said to be a *compact operator* if Ψ maps bounded sets to pre-compact sets, i.e., if given any bounded sequence $\{\phi_n\}$, there exists a subsequence so that $\Psi\phi_{n_k}$ converges to a limit. Let $\mathfrak{C}(H)$ be the set of compact operators. Ψ is said to be a *finite rank operator* if the range of Ψ is finite-dimensional. Let $\mathfrak{F}(H)$ be the set of finite rank operators.

Lemma 16.39   Let H be a Hilbert space.

1. $\mathfrak{F}(H)$ is a linear subspace of $\mathrm{Hom}(H)$.
2. $\mathfrak{C}(H)$ is a closed linear subspace of $\mathrm{Hom}(H)$.
3. $\mathfrak{F}(H)$ is a dense subset of $\mathfrak{C}(H)$.
4. If $\Psi \in \mathfrak{F}(H)$, there exists n and a sequence $\{\psi_i\}_{i=1}^n$ so that $\Psi\phi = \sum_{i=1}^n (\phi, \Psi^*\psi_i)\psi_i$.
5. If $\Psi \in \mathfrak{F}(H)$, then $\Psi^* \in \mathfrak{F}(H)$. If $\Psi \in \mathfrak{C}(H)$, then $\Psi^* \in \mathfrak{C}(H)$.
6. $\mathfrak{C}(H)$ is a closed two-sided ideal of $\mathrm{Hom}(H)$.

Proof. Throughout the proof, we shall let $\{\phi_i\}$ be a bounded sequence of elements of H, and we shall let $\kappa := \sup \|\phi_i\|$. Since $\mathrm{range}\{c_1\Psi_1 + c_2\Psi_2\} \subset \mathrm{range}\{\Psi_1\} + \mathrm{range}\{\Psi_2\}$, $\mathfrak{F}(H)$ is a linear subspace of $\mathrm{Hom}(H)$. This establishes Assertion 1.

Let $\{\Psi_\nu\}_{\nu=1}^\infty$ be a countable sequence of elements of $\mathfrak{C}(H)$. By passing to a diagonal subsequence, we may assume $\Psi_\nu \phi_i \to \psi_\nu$ for all ν. We show $\mathfrak{C}(H)$ is a linear subset of $\mathrm{Hom}(H)$ by verifying $(c_1 \Psi_1 + c_2 \Psi_2)\phi_i \to c_1 \psi_1 + c_2 \psi_2$. Suppose Ψ_ν is a sequence of elements in $\mathfrak{C}(H)$ so that $\Psi_\nu \to \Psi$ in $(\mathrm{Hom}(H), |||)$. Let $\varepsilon > 0$ be given. Choose ν_ε so that $|||\Psi - \Psi_{\nu_\varepsilon}||| < \varepsilon$. Choose N so that $\|\Psi_{\nu_\varepsilon}(\phi_i - \phi_j)\| < \varepsilon$ for $i, j > N$. If $i, j > N$,

$$
\begin{aligned}
\|\Psi(\phi_i - \phi_j)\| &\leq \|\Psi_{\nu_\varepsilon}(\phi_i - \phi_j)\| + \|(\Psi - \Psi_{\nu_\varepsilon})(\phi_i - \phi_j)\| \\
&\leq \varepsilon + |||\Psi - \Psi_{\nu_\varepsilon}||| \, \|\phi_i - \phi_j\| \leq \varepsilon + 2\varepsilon\kappa \,.
\end{aligned}
$$

Consequently, $\{\Psi\phi_i\}$ is a Cauchy sequence so $\Psi \in \mathfrak{C}(H)$ is compact. Thus, $\mathfrak{C}(H)$ is a closed linear subspace of $\mathrm{Hom}(H)$. This establishes Assertion 2.

Let $B_r(\phi) := \{\psi \in H : \|\phi - \psi\| < r\}$ be the open ball of radius r about ϕ in H. Let $\Psi \in \mathfrak{F}(H)$. Let $C := \mathrm{range}\{\Psi\}$. Since C is finite-dimensional, $K_\Psi := \overline{B_\kappa(0)} \cap C$ is compact. Since $\Psi\phi_i \in K_\Psi$, there exists a convergent subsequence of the sequence $\Psi\phi_i$. This shows that

$$
\mathfrak{F}(H) \subset \mathfrak{C}(H) \,.
$$

Let $\Psi \in \mathfrak{C}(H)$. By assumption, $K_\Psi := \overline{\Psi B_1(0)}$ is a compact subset of H. Let $\varepsilon > 0$ be given. Choose elements $\{y_1, \ldots, y_n\}$ in K_Ψ so that $K_\Psi \subset \cup_i B_{\frac{1}{2}\varepsilon}(y_i)$. Choose $\{x_1, \ldots, x_n\}$ so

$$
y_i \in B_{\frac{1}{2}\varepsilon}(\Psi x_i) \,.
$$

Consequently, $K_\Psi \subset \cup_i B_\varepsilon(\Psi x_i)$. Let $C := \mathrm{span}\{\Psi x_i\}$. Let π_C be the orthogonal projection on C of Construction 16.35; we have $|||\pi_C||| \leq 1$. If $x \in B_1(0)$, then $\Psi x \in K_\Psi$ so we may choose i with $1 \leq i \leq n$ so $\|\Psi x - \Psi x_i\| < \varepsilon$. Because $\pi_C \Psi x_i = \Psi x_i$, we may estimate

$$
\begin{aligned}
\|\Psi x - \pi_C \Psi(x)\| &\leq \|\Psi x - \Psi x_i\| + \|\Psi x_i - \pi_C \Psi x\| \\
&= \|\Psi x - \Psi x_i\| + \|\pi_C \Psi x_i - \pi_C \Psi x\| \\
&\leq \varepsilon + |||\pi_C||| \, \|\Psi x_i - \pi_C \Psi x\| = 2\varepsilon \,.
\end{aligned}
$$

This shows that $|||\Psi - \pi_C \Psi||| < 2\varepsilon$ and hence, since $\pi_C \Psi \in \mathfrak{F}(H)$, $\mathfrak{F}(H)$ is a dense subset of $\mathfrak{C}(H)$. This establishes Assertion 3.

Let $\Psi \in \mathfrak{F}(H)$. Let $\{\psi_1, \ldots, \psi_n\}$ be an orthonormal basis for $\mathrm{range}\{\Psi\}$. We establish Assertion 4 by expressing $\Psi\phi = \sum_{1 \leq i \leq n}(\Psi\phi, \psi_i)\psi_i = \sum_{1 \leq i \leq n}(\phi, \Psi^*\psi_i)\psi_i$. Let $\phi^* \in H$. We have dually

$$
(\Psi^*\phi^*, \phi) = (\phi^*, \Psi\phi) = \sum_{i=1}^n (\phi^*, \psi_i)\overline{(\phi, \Psi^*\psi_i)} = \left(\sum_{i=1}^n (\phi^*, \psi_i)\Psi^*\psi_i, \phi \right) \quad \text{so}
$$

$$
\Psi^*\phi^* = \sum_{i=1}^n (\phi^*, \psi_i)\Psi^*\psi_i \quad \text{and} \quad \mathrm{range}\{\Psi^*\} \subset \mathrm{span}\{\Psi^*\psi_i\} \,.
$$

Consequently, $\Psi^* \in \mathfrak{F}(H)$. By Lemma 16.37, the map $\Psi \to \Psi^*$ preserves $\||\cdot\||$; we complete the proof of Assertion 5 by noting $\mathfrak{C}(H)^* = \left\{\overline{\mathfrak{F}(H)}\right\}^* = \overline{\mathfrak{F}(H)^*} = \overline{\mathfrak{F}(H)} = \mathfrak{C}(H)$.

Suppose $\Psi \in \mathfrak{C}(H)$ and $\Phi \in \mathrm{Hom}(H)$. Since $\{\Phi\phi_i\}$ is a bounded sequence and Ψ is a compact operator, there exists a subsequence so $\Psi\Phi\phi_i$ converges in H. Thus, $\Psi\Phi$ belongs to $\mathfrak{C}(H)$. Since Ψ is compact, there exists a subsequence so $\Psi\phi_i$ converges. Since Φ is continuous, $\Phi\Psi\phi_i$ converges. Consequently, $\Phi\Psi$ belongs to $\mathfrak{C}(H)$. This shows that $\mathfrak{C}(H)$ is a two-sided ideal of $\mathrm{Hom}(H)$. □

16.4.3 COMPACT SELF-ADJOINT OPERATORS.

Lemma 16.40   Let H be an infinite-dimensional Hilbert space. Let $\Psi \in \mathfrak{C}(H)$ be self-adjoint.

1. If $\lambda \neq 0$, then $E(\lambda)$ is finite-dimensional.
2. Let $\{\lambda_i\}$ be distinct eigenvalues. Then $\lim_{i\to\infty} \lambda_i = 0$.
3. $E(\||\Psi\||) \oplus E(-\||\Psi\||)$ is non-trivial.

Proof. Suppose to the contrary that $E(\lambda)$ is infinite-dimensional. Since $E(\lambda)$ is closed, $E(\lambda)$ is a Hilbert space and we may use the Gram–Schmidt process of Construction 16.30 to find an orthonormal sequence $\{\phi_i\}_{i=1}^{\infty}$ in $E(\lambda)$. Since $\|\phi_i\| = 1$ and since Ψ is compact, we can find a subsequence ϕ_{i_k} so $\Psi\phi_{i_k}$ is a Cauchy sequence. Since $\Psi\phi_{i_k} = \lambda\phi_{i_k}$ and $\lambda \neq 0$, we conclude the sequence $\{\phi_{i_k}\}$ is a Cauchy sequence. This is impossible as the $\{\phi_{i_k}\}$ are an orthonormal sequence. Consequently, $\|\phi_i - \phi_j\|^2 \geq 2$ for $i \neq j$. Assertion 1 follows.

Suppose Assertion 2 fails. We argue for a contradiction and assume there exists a sequence of distinct eigenvalues $\{\lambda_i\}$ so that $|\lambda_i| \geq \varepsilon > 0$. Choose unit eigenvectors $\phi_i \in E(\lambda_i)$. By Lemma 16.38, the eigenvalues λ_i are real and $\phi_i \perp \phi_j$. Since Ψ is compact, by passing to a subsequence if necessary, we can assume $\{\Psi\phi_i\} = \{\lambda_i\phi_i\}$ is a Cauchy sequence. Let $i \neq j$. Since $\phi_i \perp \phi_j$, $\|\lambda_i\phi_i - \lambda_j\phi_j\|^2 = \lambda_i^2 + \lambda_j^2$. Since this tends to zero, we have $\lambda_i \to 0$. This contradiction establishes Assertion 2.

If $\lambda := \||\Psi\|| = 0$, then $\Psi = 0$ and Assertion 3 is trivial. So we assume $\lambda \neq 0$. Choose a sequence of unit vectors ϕ_n so $\|\Psi\phi_n\| \to \lambda$. As Ψ is compact, by passing to a subsequence if necessary, we can assume that $\Psi\phi_n \to \phi_\infty$. We compute

$$
\begin{aligned}
\|\Psi^2\phi_n - \lambda^2\phi_n\|^2 &= \|\Psi^2\phi_n\|^2 + \lambda^4\|\phi_n\|^2 - \lambda^2(\Psi^2\phi_n, \phi_n) - \lambda^2(\phi_n, \Psi^2\phi_n) \\
&= \|\Psi^2\phi_n\|^2 + \lambda^4\|\phi_n\|^2 - 2\lambda^2(\Psi\phi_n, \Psi\phi_n) \\
&\leq \lambda^4 + \lambda^4 - 2\lambda^2\|\Psi\phi_n\|^2 \to 0.
\end{aligned}
$$

Since $\Psi\phi_n \to \phi_\infty$, $\Psi^2\phi_n \to \Psi\phi_\infty$. Thus, $\lambda^2\phi_n \to \Psi\phi_\infty$. Since $\lambda^2 \neq 0$, $\phi_n \to \lambda^{-2}\Psi\phi_\infty$. Because ϕ_n is a unit vector, $\|\lambda^{-2}\Psi\phi_\infty\| = 1$. In particular, $\phi_\infty \neq 0$. Since $\Psi\phi_n \to \phi_\infty$, $\phi_\infty = \lambda^{-2}\Psi^2\phi_\infty$, and thus $0 = (\Psi^2 - \lambda^2)\phi_\infty = (\Psi + \lambda)(\Psi - \lambda)\phi_\infty$. If $(\Psi - \lambda)\phi_\infty = 0$, then $E(\lambda) \neq 0$. On the other hand, if $(\Psi - \lambda)\phi_\infty \neq 0$, then $E(-\lambda) \neq 0$. □

The following result generalizes the standard spectral theory in finite dimensions to the infinite-dimensional context.

Lemma 16.41   Let H be an infinite-dimensional separable Hilbert space. Let Ψ be compact and self-adjoint. There exists a complete orthonormal basis $\{\phi_n\}_{n=1}^{\infty}$ for H so that $\Psi\phi_n = \lambda_n\phi_n$ where $|\lambda_1| \geq |\lambda_2| \geq \cdots \geq 0$, and $\lim_{n\to\infty}\lambda_n = 0$; the collection $\{\phi_n, \lambda_n\}$ is called a *complete spectral resolution* of Ψ.

Proof. The result is immediate if $\Psi = 0$ so we may assume without loss of generality that $\Psi \neq 0$. By Lemma 16.40, the space $E(\||\Psi\||) \oplus E(-\||\Psi\||)$ is non-trivial and finite-dimensional. Let $H_1 := \{E(\||\Psi\||) \oplus E(-\||\Psi\||)\}^{\perp}$. Since Ψ is self-adjoint, H_1 is invariant under Ψ. Let Ψ_1 be the restriction of Ψ to H_1; $\||\Psi_1\||_{H_1} \leq \||\Psi\||$. If equality holds, then we can find a non-trivial eigenvector with eigenvalue $\pm\||\Psi\||$ in H_1 by Lemma 16.40; this is false as we have already factored out those eigenvalues. Thus, $\||\Psi_1\||_{H_1} < \||\Psi\||_{H_1}$. We can continue in this way to build a decreasing positive sequence $\{\lambda_n\}$ so $E(\lambda_n) \oplus E(-\lambda_n) \neq \{0\}$ and so that if Ψ_n is the restriction of Ψ to $H_n := \{E(\lambda_1) \oplus \cdots \oplus E(\lambda_n) \oplus E(-\lambda_1) \oplus \cdots \oplus E(-\lambda_n)\}^{\perp}$, then $\||\Psi_n\||_{H_n} < \lambda_n$. If the sequence terminates, then we are done as $\Psi_{n+1} = 0$. Otherwise, we build an infinite sequence of eigenvectors so by Lemma 16.40, $|\lambda_n| \to 0$. Taking the limit and setting $H_{\infty} = \cap_n H_n$, we obtain $\Psi_{\infty} = 0$. Choosing an orthonormal bases for $E(\pm\lambda_n)$ and $H_{\infty} = E(0)$ constructs the desired orthonormal basis for H.   □

16.4.4 FREDHOLM OPERATORS.

Let $\Psi \in \mathrm{Hom}(H_1, H_2)$ where H_1 and H_2 are Hilbert spaces. Let $\mathfrak{K}_{\Psi} := \ker\{\Psi\}$, $\mathfrak{R}_{\Psi} := \mathrm{range}\{\Psi\}$, and $\mathfrak{C}_{\Psi} = \mathrm{coker}\{\Psi\} := H_2/\mathfrak{R}_{\Psi}$. Ψ is said to be *Fredholm* if $\mathfrak{K}_{\Psi}$ and $\mathfrak{C}_{\Psi}$ are finite-dimensional. If Ψ is Fredholm, set

$$\mathrm{Index}(\Psi) := \dim(\mathfrak{K}_{\Psi}) - \dim(\mathfrak{C}_{\Psi}).$$

Let $\mathrm{Fred}(H_1, H_2)$ be the set of all Fredholm operators; the notation is chosen in honor of the seminal paper by Fredholm [23]. If H is a separable Hilbert space, we apply Lemma 16.39 to see $\mathfrak{C}(H)$ is a closed two-sided ideal of $\mathrm{Hom}(H)$. The quotient $\mathrm{Hom}(H)/\mathfrak{C}(H)$ is called the *Calkin algebra* [13]. The operators in $\mathrm{Hom}(H)$ which project to the invertible elements in the Calkin algebra are the Fredholm operators.

Let $\Psi \in \mathrm{Fred}(H_1, H_2)$. Since $\mathfrak{K}_{\Psi}$ is finite-dimensional, $\mathfrak{K}_{\Psi}$ is a closed linear subspace of H_1. Let $\pi_{\mathfrak{K}_{\Psi}}$ be orthogonal projection from H_1 to $\mathfrak{K}_{\Psi}$; $\||\pi_{\mathfrak{K}_{\Psi}}\|| = 1$. Give $\mathfrak{K}_{\Psi}$ the induced Hilbert space structure. Let $\pi_{\mathfrak{C}_{\Psi}}$ be the natural projection from H_2 to $\mathfrak{C}_{\Psi}$. Since $\mathfrak{C}_{\Psi}$ is finite-dimensional, we can use the Gram–Schmidt process to choose a finite orthonormal subset $\{\psi_i\}$ of H_2 so that $\{\pi_{\mathfrak{C}_{\Psi}}\psi_i\}$ is a basis for $\mathfrak{C}_{\Psi}$. Give $\mathfrak{C}_{\Psi}$ a Hilbert space structure by requiring that this is an orthonormal basis. Define a splitting ι_{Ψ} of $\pi_{\mathfrak{C}_{\Psi}}$ by requiring $\iota_{\Psi}(\pi_{\mathfrak{C}_{\Psi}}\psi_i) = \psi_i$. Since $\mathfrak{C}_{\Psi}$ is finite-dimensional, ι_{Ψ} is continuous. Let $\tilde{H}_1 := H_1 \oplus \mathfrak{C}_{\Psi}$ and $\tilde{H}_2 = H_2 \oplus \mathfrak{K}_{\Psi}$. If Φ belongs to $\mathrm{Hom}(H_1, H_2)$, define $\Theta_{\Psi}(\Phi) \in \mathrm{Hom}(\tilde{H}_1, \tilde{H}_2)$ by

$$\Theta_{\Psi}(\Phi)(\phi \oplus c) := (\Phi\phi + \iota_{\Psi}c) \oplus \pi_{\mathfrak{K}_{\Psi}}\phi. \qquad (16.4.a)$$

Let $i_1(\psi) := \psi \oplus 0$ be the natural inclusion of H_1 in $\tilde{H}_1$ and let $\pi_1 : \tilde{H}_2 \to H_2$ be projection on the first factor. We may then express

$$\Phi = \pi_1 \circ \Theta_\Psi(\Phi) \circ i_1.\qquad(16.4.b)$$

The map $\Theta_\Psi : \mathrm{Hom}(H_1, H_2) \to \mathrm{Hom}(\tilde{H}_2, \tilde{H}_2)$ is not a linear map. However,

$$\{\Theta_\Psi(\Phi_1) - \Theta_\Psi(\Phi_2)\}(\phi \oplus c) = (\Phi_1 - \Phi_2)\phi \oplus 0.$$

Consequently, $\||\Theta_\Psi\Phi_1 - \Theta_\Psi\Phi_2\|| = \||\Phi_1 - \Phi_2\||$. This shows that Θ_Ψ is norm preserving and hence continuous.

Lemma 16.42   Let H_1, H_2, and H_3 be infinite-dimensional separable Hilbert spaces, let Ψ belong to $\mathrm{Fred}(H_1, H_2)$, and let Φ belong to $\mathrm{Fred}(H_2, H_3)$.

1. $\mathrm{range}\{\Psi\}$ is closed.
2. $\Phi \circ \Psi \in \mathrm{Fred}(H_1, H_3)$.
3. $\mathrm{Index}(\Phi \circ \Psi) = \mathrm{Index}(\Phi) + \mathrm{Index}(\Psi)$.
4. $\mathrm{Fred}(H_1, H_2)$ is an open subset of $\mathrm{Hom}(H_1, H_2)$.
5. Index is a continuous function from $\mathrm{Fred}(H_1, H_2)$ to $\mathbb{Z}$.

Proof. We adopt the notation of Section 16.4.4. It is an easy algebraic exercise to verify $\Theta_\Psi(\Psi)$ is 1-1 and onto. Consequently, by Theorem 16.20, $\Theta_\Psi(\Psi)^{-1}$ is a continuous linear map from $\tilde{H}_2$ to $\tilde{H}_1$. Let $\psi \in \overline{\mathfrak{R}_\Psi}$. We must show $\psi \in \mathfrak{R}_\Psi$. Choose $\phi_n \in H_1$ so $\Psi\phi_n \to \psi \in H_2$. Replace ϕ_n by $\tilde{\phi}_n := \phi_n - \pi_{\mathfrak{K}_\Psi}\phi_n$. Then $\pi_{\mathfrak{K}_\Psi}\tilde{\phi}_n = 0$ so $\Psi\tilde{\phi}_n = \Psi\phi_n \to \psi$. We have

$$\Theta_\Psi(\Psi)(\tilde{\phi}_n \oplus 0) = (\Psi\tilde{\phi}_n, 0) \to (\psi \oplus 0).$$

Applying $\Theta_\Psi(\Psi)^{-1}$ yields

$$(\tilde{\phi}_n \oplus 0) = \Theta_\Psi(\Psi)^{-1}(\Psi\tilde{\phi}_n, 0) \to \Theta_\Psi(\Psi)^{-1}(\psi \oplus 0).$$

Set $\phi_\infty := \Theta_\Psi(\Psi)^{-1}(\psi \oplus 0)$. Then $\Psi\tilde{\phi}_n \to \psi$ and $\Psi\tilde{\phi}_n \to \Psi\phi_\infty$ implies $\Psi\phi_\infty = \psi$ so ψ belongs to $\mathfrak{R}_\Psi$. This establishes Assertion 1.

By Assertion 1, $\mathfrak{R}_\Psi$ is closed. Decompose $H_2 = \mathfrak{R}_\Psi \oplus \mathfrak{R}_\Psi^\perp$. Projection on the second factor identifies $\mathfrak{C}_\Psi$ with $\mathfrak{R}_\Psi^\perp$ and we replace the original splitting ι_Ψ discussed in Section 16.4.4 with this isometric splitting. We have $\mathfrak{R}_\Psi = \Psi(\mathfrak{K}_\Psi^\perp)$ and $\mathfrak{C}_\Psi = \mathfrak{R}_\Psi^\perp = \Psi(\mathfrak{K}_\Psi^\perp)^\perp$. Decompose

$$
\begin{aligned}
H_1 &= \mathfrak{K}_\Psi \oplus \mathfrak{K}_\Psi^\perp, & H_2 &= \mathfrak{K}_\Phi \oplus \mathfrak{K}_\Phi^\perp = \Psi(\mathfrak{K}_\Psi^\perp) \oplus \Psi(\mathfrak{K}_\Psi^\perp)^\perp, \\
H_3 &= \Phi(\mathfrak{K}_\Phi^\perp) \oplus \Phi(\mathfrak{K}_\Phi^\perp)^\perp = \Phi(\mathfrak{K}_\Phi^\perp \cap \Psi(\mathfrak{K}_\Psi^\perp)) \oplus \Phi(\mathfrak{K}_\Phi^\perp \cap \Psi(\mathfrak{K}_\Psi^\perp)^\perp) \oplus \Phi(\mathfrak{K}_\Phi^\perp)^\perp, & &(16.4.c) \\
\mathfrak{K}_{\Phi \circ \Psi} &= \mathfrak{K}_\Psi \oplus \mathfrak{K}_\Phi \cap \Psi(\mathfrak{K}_\Psi^\perp), & \mathfrak{C}_{\Phi \circ \Psi} &= \Phi(\mathfrak{K}_\Phi^\perp \cap \Psi(\mathfrak{K}_\Psi^\perp)^\perp) \oplus \Phi(\mathfrak{K}_\Phi^\perp)^\perp.
\end{aligned}
$$

This shows that $\mathfrak{K}_{\Psi \circ \Psi}$ and $\mathfrak{C}_{\Phi \circ \Psi}$ are finite-dimensional. Consequently, $\Phi \circ \Psi$ is Fredholm; this proves Assertion 2.

We use Equation (16.4.c) to establish Assertion 3 by computing:

$\mathrm{Index}(\Phi \circ \Psi)$

$$= \dim(\mathfrak{K}_\Psi) + \dim(\mathfrak{K}_\Phi \cap \Psi(\mathfrak{K}_\Psi^\perp)) - \dim(\Phi(\mathfrak{K}_\Phi^\perp \cap \Psi(\mathfrak{K}_\Psi^\perp))^\perp) - \dim(\Phi(\mathfrak{K}_\Phi^\perp)^\perp)$$

$$= \dim(\mathfrak{K}_\Psi) + \dim(\mathfrak{K}_\Phi \cap \Psi(\mathfrak{K}_\Psi^\perp)) + \dim(\mathfrak{K}_\Phi \cap \Psi(\mathfrak{K}_\Psi^\perp)^\perp)$$

$$\quad - \dim(\mathfrak{K}_\Phi \cap \Psi(\mathfrak{K}_\Psi^\perp)^\perp) - \dim(\Phi(\mathfrak{K}_\Phi^\perp \cap \Psi(\mathfrak{K}_\Psi^\perp)^\perp)) - \dim(\Phi(\mathfrak{K}_\Phi^\perp)^\perp)$$

$$= \dim(\mathfrak{K}_\Psi) + \dim(\mathfrak{K}_\Phi) - \dim(\Psi(\mathfrak{K}_\Psi^\perp)^\perp) - \dim(\Phi(\mathfrak{K}_\Phi^\perp)^\perp)$$

$$= \mathrm{Index}(\Psi) + \mathrm{Index}(\Phi) \,.$$

Any two separable Hilbert spaces are isomorphic so we may set $H := \tilde{H}_1 = \tilde{H}_2$ in the proof of the remaining two assertions. Since $\Theta_\Psi(\Psi)$ is 1-1 and onto, $\Theta_\Psi(\Psi) \in \mathrm{GL}(H)$. By Lemma 16.21, $\mathrm{GL}(H)$ is an open subset of $\mathrm{Hom}(H)$. Since Θ is a continuous map, there exists $\varepsilon > 0$ so that if $\|\|\Phi - \Psi\|\| < \varepsilon$, then $\Theta_\Psi(\Phi) \in \mathrm{GL}(\tilde{H})$. Thus, $\Theta_\Psi(\Phi)$ is Fredholm. By Equation (16.4.b), $\Phi = \pi_1 \circ \Theta_\Psi(\Phi) \circ i_1$. It is immediate from the definition that π_1 and i_1 are Fredholm. Consequently, by Assertion 2, Φ is Fredholm. This proves Assertion 4. By Assertion 3, $\mathrm{Index}(\Phi) = \mathrm{Index}(\pi_1) + \mathrm{Index}(\Theta_\Psi(\Phi)) + \mathrm{Index}(i_1)$. Since $\Theta_\Psi(\Phi) \in \mathrm{GL}(\tilde{H})$, $\mathrm{Index}(\Theta(\Phi)) = 0$. This shows $\mathrm{Index}(\Phi)$ is locally constant and Assertion 5 follows. $\qquad\square$

Elliptic Operator Theory

We shall discuss work of the following mathematicians, among others, in Chapter 17:

W. Allard

S. Bochner
(1899–1982)

P. Dirac
(1902–1984)

J. Fourier
(1768–1830)

L. Gårding
(1919–2014)

W. Hodge
(1903–1975)

P. Laplace
(1749–1827)

M. Plancherel
(1885–1967)

F. Rellich
(1906–1955)

L. Schwartz
(1915–2002)

S. Sobolev
(1908–1989)

In Section 17.1, we describe the Fourier transform; this is a fundamental tool in the study of partial differential equations. In Section 17.2, we define the Sobolev norms on the Schwarz space in $\mathbb{R}^m$ and examine various properties of these norms. We then turn our attention to the setting of compact Riemannian manifolds. In Section 17.3, we define several norms on the space of smooth sections to a vector bundle. We also examine the geometry of operators of Laplace type. We conclude Chapter 17 in Section 17.4 by discussing the spectral theory of a self-adjoint operator of Laplace type; this is the main result of Chapter 17 and forms the foundation of our treatment of various problems in Differential Geometry in the remaining chapters.

Throughout Chapter 17, C will be a generic finite positive constant; ocasionally, it will be important to indicate the parameters upon which it depends, but for the most part we shall not bother in the interests of notational simplicity. Let (V, h, ∇_V) be a *geometric vector bundle* over a compact Riemannian manifold (M, g) of dimension m throughout this chapter. Here, V is a smooth vector bundle over M which is equipped with a Hermitian fiber metric h and a connection ∇_V. We shall usually assume V is complex; the corresponding real setting is handled analogously. We use (g, h) to define fiber metrics on tensors of all types and the Levi–Civita connection ∇^g and the connection ∇_V to covariantly differentiate tensors of all types. Let $\vec{s} = (s^1, \ldots, s^\ell)$ be a local frame for V. By applying the Gram–Schmidt process, we can always assume that this is a local orthonormal frame. Let $\vec{x} = (x^1, \ldots, x^m)$ be a system of local coordinates on M.

17.1   FOURIER TRANSFORM

In Section 17.1.1, we introduce the Schwarz space $\mathcal{S}$. This is not a Banach space, but rather is a Fréchet space; the topology is defined by a countable sequence of increasing norms. In Section 17.1.2, we introduce the Fourier transform; the natural domain is the Schwarz space as the Fourier inversion formula, which we establish in Section 17.1.3, shows the Fourier transform is a homeomorphism of $\mathcal{S}$. In Section 17.1.4, we will establish the Plancherel formula which shows that the Fourier transform extends to an isometry of $L^2(\mathbb{R}^m)$ (with a suitably normalized measure). The Schwarz space has two different commutative ring structures. The first is given by function multiplication and the second is given by convolution. In Section 17.1.5, we show the Fourier transform interchanges these two structures.

Let $x = (x^1, \ldots, x^m)$ belong to $\mathbb{R}^m$ and let $\xi = (\xi_1, \ldots, \xi_m)$ belong to the dual space $(\mathbb{R}^m)^*$. Let $dx = dx^1 \cdots dx^m$ be the measure defined by the usual flat metric on $\mathbb{R}^m$. We set

$$\partial_x^\alpha := (\partial_{x^1})^{a_1} \ldots (\partial_{x^m})^{a_m}, \quad x^\alpha := (x^1)^{a_1} \ldots (x^m)^{a_m},$$
$$x \cdot \xi = x^1 \xi_1 + \cdots + x^m \xi_m, \quad |\alpha| = a_1 + \cdots + a_m,$$

where $\alpha = (a_1, \ldots, a_m)$ is a multi-index. If $f \in C^\infty(\mathbb{R}^m)$, we may define the *sup norm* $\|\cdot\|_\infty$ and the L^p *norm* $\|\cdot\|_{L^p}$ of f by setting:

$$\|f\|_\infty := \sup_{x \in \mathbb{R}^m} |f(x)| \quad \text{and} \quad \|f\|_{L^p} := \left\{ \int_{\mathbb{R}^m} |f|^p dx \right\}^{1/p}.$$

These can, of course, be infinite. They are finite on the space $C_0^\infty(\mathbb{R}^m)$ of smooth functions with compact support and we let $L^p(\mathbb{R}^m)$ be the completion of $C_0^\infty(\mathbb{R}^m)$ with respect to the norm $\|_{L^p}$.

17.1.1 SCHWARZ SPACE $\mathcal{S}$. This is the subspace of $C^\infty(\mathbb{R}^m)$ consisting of all smooth functions $f : \mathbb{R}^m \to \mathbb{C}$ so that f and all its derivatives decay rapidly at infinity, i.e., $\|x^\alpha \partial_x^\beta f\|_\infty < \infty$ for all α and β. Define a Fréchet space topology on $\mathcal{S}$ by introducing the following sequence of increasing norms:

$$\|f\|_j^{\mathcal{S}} := \sum_{|\alpha|+|\beta|\leq j} \|x^\alpha \partial_x^\beta f\|_\infty \,. \tag{17.1.a}$$

The following result is immediate; it rests upon Lemma 16.19 which we use to extend linear maps from a dense subset to the appropriate completions.

Lemma 17.1   Let $f \in \mathcal{S}$ and $\vartheta \in \mathcal{S}$.

1. The maps $f \to x^\alpha f$ and $f \to \partial_x^\alpha f$ are continuous linear maps from $\mathcal{S}$ to $\mathcal{S}$.
2. The map $m_\vartheta : f \to \vartheta f$ is a continuous linear map from $\mathcal{S}$ to $\mathcal{S}$.
3. $|x^\alpha \partial_x^\beta f(x)| \leq C \|f\|_{|\alpha|+|\beta|+2n}^{\mathcal{S}} (1 + \|x\|^2)^{-n}$.
4. $x^\alpha \partial_x^\beta f \in L^p(\mathbb{R}^m)$ for $1 \leq p < \infty$.
5. $\|f\|_j^{\mathcal{S}}$ and $\sum_{|\alpha|+|\beta|\leq j} \|\partial_x^\beta x^\alpha f\|_\infty$ are equivalent norms.

17.1.2 THE FOURIER TRANSFORM [21]. If $f \in \mathcal{S}$, let

$$\hat{f}(\xi) = (2\pi)^{-\frac{1}{2}m} \int_{\mathbb{R}^m} e^{-\sqrt{-1}\,x\cdot\xi} f(x) dx \,;$$

this integral, which defines the Fourier transform, is well defined by Lemma 17.1 (4). The following result shows that $\hat{f} \in \mathcal{S}$.

Lemma 17.2   Let $f \in \mathcal{S}$.

1. $\xi_i \hat{f} = \sqrt{-1}\,\widehat{\partial_{x^i} f}$ and $\partial_{\xi_i} \hat{f}(\xi) = -\sqrt{-1}\,\widehat{x^i f}$.
2. $f \to \hat{f}$ is a continuous map from $\mathcal{S}$ to $\mathcal{S}$.

Proof. The following computation establishes Assertion 1; Lemma 17.1 justifies interchanging differentiation and integration in the following integrals:

$$\partial_{\xi_i} \hat{f}(\xi) = (2\pi)^{-\frac{1}{2}m} \partial_{\xi_i} \int_{\mathbb{R}^m} e^{-\sqrt{-1}\, x\cdot\xi} f(x) dx = (2\pi)^{-\frac{1}{2}m} \int_{\mathbb{R}^m} \partial_{\xi_i} \left\{ e^{-\sqrt{-1}\, x\cdot\xi} f(x) \right\} dx$$

$$= -\sqrt{-1}\,(2\pi)^{-\frac{1}{2}m} \int_{\mathbb{R}^m} e^{-\sqrt{-1}\, x\cdot\xi} x^i f(x) dx = -\sqrt{-1}\, \widehat{x^i f}(\xi),$$

$$\xi_i \hat{f}(\xi) = (2\pi)^{-\frac{1}{2}m} \int_{\mathbb{R}^m} \xi_i e^{-\sqrt{-1}\, x\cdot\xi} f(x) dx$$

$$= \sqrt{-1}\,(2\pi)^{-\frac{1}{2}m} \int_{\mathbb{R}^m} \partial_{x^i} \left\{ e^{-\sqrt{-1}\, x\cdot\xi} \right\} f(x) dx$$

$$= -\sqrt{-1}\,(2\pi)^{-\frac{1}{2}m} \int_{\mathbb{R}^m} e^{-\sqrt{-1}\, x\cdot\xi} \partial_{x^i} f dx = -\sqrt{-1}\, \widehat{\partial_{x^i} f}.$$

Let $n > \frac{1}{2}m$ so that $(1 + \|x\|^2)^{-n} \in L^1(\mathbb{R}^m)$. By Lemma 17.1 (3), we have that

$$|\partial_\xi^\alpha \xi^\beta \hat{f}(\xi)| = |\widehat{x^\alpha \partial_x^\beta f}(\xi)| \le (2\pi)^{-\frac{1}{2}m} \int |x^\alpha \partial_x^\beta f(x)| dx$$

$$\le C\|f\|_{|\alpha|+|\beta|+n}^{\mathcal{S}} \int (1 + \|x\|^2)^{-n} dx \le C_1 \|f\|_{|\alpha|+|\beta|+n}^{\mathcal{S}}.$$

Lemma 17.1 permits us to interchange ∂_ξ^α and ξ^β and obtain $\|\hat{f}\|_j^{\mathcal{S}} \le C_2 \|f\|_{j+2n}^{\mathcal{S}}$; Assertion 2 now follows.   □

17.1.3 THE FOURIER INVERSION FORMULA.

Lemma 17.3

1. If $f \in \mathcal{S}$, then $f(x) = (2\pi)^{-\frac{1}{2}m} \int_{\mathbb{R}^m} e^{\sqrt{-1}\, x\cdot\xi} \hat{f}(\xi) d\xi$.

2. The map $f \to \hat{f}$ is a homeomorphism of $\mathcal{S}$.

Proof. If $f \in \mathcal{S}$, let

$$\mathcal{U}(f)(x) := (2\pi)^{-\frac{1}{2}m} \int_{\mathbb{R}^m} e^{\sqrt{-1}\, x\cdot\xi} \hat{f}(\xi) d\xi.$$

By Lemma 17.2, $\hat{f} \in \mathcal{S}$, so $\mathcal{U}(f) \in \mathcal{S}$ is well defined. We must show $\mathcal{U}(\hat{f}) = f$. We decompose the proof into various parts.

Step 1. Let $f_0(x) := e^{-\frac{1}{2}\|x\|^2} \in \mathcal{S}$ be the *Gaussian distribution*. We compute

$$\hat{f_0}(\xi) = (2\pi)^{-\frac{1}{2}m} \int_{\mathbb{R}^m} e^{-\frac{1}{2}x\cdot x - \sqrt{-1}\, x\cdot\xi} dx$$

$$= e^{-\frac{1}{2}\|\xi\|^2} (2\pi)^{-\frac{1}{2}m} \int_{\mathbb{R}^m} e^{-\frac{1}{2}(x+\sqrt{-1}\,\xi)\cdot(x+\sqrt{-1}\,\xi)} dx.$$

Change variables to replace x by $x + \sqrt{-1}\,\xi$ and use Green's Theorem to shift the resulting contour integral in $\mathbb{C}^m$ back to the original contour in $\mathbb{R}^m$ to obtain

$$\hat{f}_0(\xi) = e^{-\frac{1}{2}\|\xi\|^2}(2\pi)^{-\frac{1}{2}m}\int_{\mathbb{R}^m} e^{-\frac{1}{2}\|x\|^2}dx = e^{-\frac{1}{2}\|\xi\|^2}.$$

A similar argument now shows $\mathcal{U}(f_0) = f_0$ and establishes the lemma for $f = f_0$.

Step 2. Suppose $f(0) = 0$. We expand

$$f(x) = f(0) + \int_{t=0}^1 \partial_t f(tx)dt = \sum_{j=1}^m x^j \int_{t=0}^1 \partial_{x^j} f(tx)dt.$$

Set $g_j(x) := \int_{t=0}^1 \{\partial_{x^j} f\}(tx) \in \mathcal{S}$. We use Lemma 17.2 and integrate by parts to show $\mathcal{U}(f)(0) = 0$ by computing:

$$\hat{f}(\xi) \;=\; \sum_{j=1}^m \widehat{x^i g_i} = -\sqrt{-1}\sum_{i=1}^m \partial_{\xi_i}\hat{g}_i,$$

$$\mathcal{U}(f)(0) \;=\; -\sqrt{-1}\,(2\pi)^{-\frac{1}{2}m}\sum_{i=1}^m \int_{\mathbb{R}^m}(\partial_{\xi_i}\hat{g}_i)(\xi)d\xi = 0.$$

Step 3. Suppose $f(0) \neq 0$. Decompose $f = f(0)f_0 + \{f - f(0)f_0\}$. We apply Step 1 and Step 2 to see $\{\mathcal{U}(f)\}(0) = f(0)$ by computing

$$\mathcal{U}(f)(0) = f(0)\{\mathcal{U}(f_0)\}(0) + \{\mathcal{U}(f - f(0)f_0)\}(0) = f(0)f_0(0) + 0 = f(0).$$

Step 4. Let $x_0 \in \mathbb{R}^m$. Set $g(x) = f(x + x_0)$. By Step 3, $\{\mathcal{U}(g)\}(0) = g(0) = f(x_0)$. Then

$$\hat{g}(\xi) = (2\pi)^{-\frac{1}{2}m}\int_{\mathbb{R}^m} e^{-\sqrt{-1}\,x\cdot\xi} f(x + x_0)dx$$

$$= e^{\sqrt{-1}x_0\cdot\xi}(2\pi)^{-\frac{1}{2}m}\int_{\mathbb{R}^m} e^{-\sqrt{-1}(x+x_0)\cdot\xi} f(x + x_0)dx = e^{\sqrt{-1}x_0\cdot\xi}\hat{f}(\xi),$$

$$f(x_0) = \{\mathcal{U}(g)\}(0) = (2\pi)^{-\frac{1}{2}m}\int_{\mathbb{R}^m}\hat{g}(\xi)d\xi = (2\pi)^{-\frac{1}{2}m}\int_{\mathbb{R}^m} e^{\sqrt{-1}x_0\cdot\xi}\hat{f}(\xi)d\xi$$

$$= \mathcal{U}(f)(x_0).$$

This proves Assertion 1. Assertion 2 follows from Assertion 1. □

17.1.4 THE PLANCHEREL FORMULA [58].

Lemma 17.4

1. If $f, g \in \mathcal{S}$, then $(f, g)_{L^2} = (\hat{f}, \hat{g})_{L^2}$.
2. The Fourier transform extends to an isometry of $L^2(\mathbb{R}^m)$.

Proof. As the integrals in question are absolutely convergent, we may interchange the order of integration and use the Fourier inversion formula to see that

$$(\hat{f}, \hat{g})_{L^2} = (2\pi)^{-\frac{1}{2}m} \int_{\mathbb{R}^m} \int_{\mathbb{R}^m} \overline{\hat{g}(\xi)} f(x) e^{-\sqrt{-1}x\cdot\xi} dx d\xi$$

$$= (2\pi)^{-\frac{1}{2}m} \int_{\mathbb{R}^m} \int_{\mathbb{R}^m} \overline{\hat{g}(\xi)} f(x) e^{-\sqrt{-1}x\cdot\xi} d\xi dx$$

$$= (2\pi)^{-\frac{1}{2}m} \int_{\mathbb{R}^m} \overline{\left\{ \int_{\mathbb{R}^m} \hat{g}(\xi) e^{\sqrt{-1}x\cdot\xi} d\xi \right\}} f(x) dx = \int_{\mathbb{R}^m} \bar{g}(x) f(x) dx$$

$$= (f, g)_{L^2} .$$

This proves Assertion 1; Assertion 2 follows since $\mathcal{S}$ is dense in $L^2(\mathbb{R}^m)$.   □

17.1.5 CONVOLUTION. Let $f \in \mathcal{S}$ and $g \in \mathcal{S}$. The convolution $f \star g \in \mathcal{S}$ is given by

$$(f \star g)(x) := (2\pi)^{-\frac{1}{2}m} \int_{\mathbb{R}^m} f(x - y) g(y) dy .$$

We shall omit the proof of the following lemma as it is straightforward.

Lemma 17.5

1. Convolution is a continuous map from $\mathcal{S} \times \mathcal{S}$ to $\mathcal{S}$.
2. $f \star g = g \star f$ and $(f \star g) \star h = f \star (g \star h)$.
3. Convolution gives $\mathcal{S}$ the structure of a commutative algebra.

We can give $\mathcal{S}$ a different structure by using pointwise multiplication by setting

$$(f \cdot g)(x) := f(x) g(x) .$$

The Fourier transform intertwines these two ring structures.

Lemma 17.6   Let $f \in \mathcal{S}$ and $g \in \mathcal{S}$. Then $\hat{f} \cdot \hat{g} = \widehat{f \star g}$ and $\hat{f} \star \hat{g} = \widehat{f \cdot g}$.

Proof. Set $x = z - y$. We change variables to see:

$$(\hat{f} \cdot \hat{g})(\xi) = (2\pi)^{-m} \int_{\mathbb{R}^{2m}} f(x) e^{\sqrt{-1}x \cdot \xi} g(y) e^{\sqrt{-1}y \cdot \xi} dx dy$$

$$= (2\pi)^{-m} \int_{\mathbb{R}^{2m}} f(z - y) e^{\sqrt{-1}(z-y) \cdot \xi} g(y) e^{\sqrt{-1}y \cdot \xi} dz dy \qquad (17.1.\text{b})$$

$$= (2\pi)^{-m} \int_{\mathbb{R}^{2m}} f(z - y) e^{\sqrt{-1}z \cdot \xi} g(y) dz dy \,.$$

Let $n > \frac{1}{2}m$. By Lemma 17.1,

$$|g(y)| \leq C(1 + \|y\|^2)^{-2n} \quad \text{and} \quad |f(z - y)| \leq C(1 + \|z - y\|^2)^{-n}.$$

By Lemma 16.27, $(1 + \|z - y\|^2)^{-n} \leq C_1(1 + \|z\|^2)^{-n}(1 + \|y\|^2)^n$. Consequently,

$$|f(z - y)g(y)| \quad \leq \quad C_2(1 + \|z - y\|^2)^{-n}(1 + \|y\|^2)^{n-2n} \leq C_3(1 + \|z\|^2)^{-n}(1 + \|y\|^2)^{-n} \,.$$

Since $(1 + \|y\|^2)^{-n} (1 + \|z\|^2)^{-n}$ is integrable, the integrals of Equation (17.1.b) are absolutely convergent. We may therefore apply Fubini's Theorem to interchange the order of integration and compute

$$(\hat{f} \cdot \hat{g})(\xi) \quad = \quad (2\pi)^{-m} \int_{\mathbb{R}^{2m}} f(z - y) e^{\sqrt{-1}z \cdot \xi} g(y) dy dz$$

$$= \quad (2\pi)^{-\frac{1}{2}m} \int_{\mathbb{R}^m} (f \star g)(z) e^{\sqrt{-1}z \cdot \xi} dz = (\widehat{f \star g})(\xi) \,.$$

We replace f by $\hat{f}$ and g by $\hat{g}$ to see $(\widehat{\hat{f}} \cdot \widehat{\hat{g}})(\xi) = (\widehat{\hat{f} \star \hat{g}})(\xi)$. By Lemma 17.3, $\widehat{\hat{f}}(\xi) = f(-\xi)$ and $\widehat{\hat{g}}(\xi) = g(-\xi)$. Consequently, we obtain $(f \cdot g)(-\xi) = (\widehat{\hat{f} \star \hat{g}})(\xi)$. We take the Fourier transform to see $(\widehat{f \cdot g})(-\xi) = (\hat{f} \star \hat{g})(\xi)$. We apply Lemma 17.3 once again to complete the proof by verifying $(\widehat{f \cdot g})(-\xi) = (\hat{f} \star \hat{g})(-\xi)$. □

Let $B_r := \{\vec{x} : \|x\| < r\}$ be the open ball of radius r about the origin. We will need the following technical result in Section 17.3.2.

Lemma 17.7   Let g_n be a sequence of non-negative smooth functions with compact support in $B_{\frac{1}{n}}$ with $\int_{\mathbb{R}^m} g_n = 1$. Let f be a C^k function with compact support in B_r. Set $f_n := f \star g_n$.

1. f_n is a smooth function with compact support in $B_{r + \frac{1}{n}}$.
2. If $|\alpha| \leq k$, then $\partial_x^\alpha f_n$ converges uniformly to $\partial_x^\alpha f$.

Proof. We have that

$$f_n(x) = \int_{\mathbb{R}^m} g_n(x - y) f(y) dy = \int_{\mathbb{R}^m} g_n(y) f(x - y) dy \,.$$

As the integral defining f_n converges uniformly and the domain of integration is in fact compact, we can differentiate under the integral sign to conclude f_n is smooth:

$$\{\partial_x^\alpha f_n\}(x) = \partial_x^\alpha \int_{\mathbb{R}^m} g_n(x-y) f(y) dy = \int_{\mathbb{R}^m} \partial_x^\alpha g_n(x-y) f(y) dy.$$

Since f has support in the ball of radius r, the integral ranges over $\|y\| \le r$. As g_n has support in the ball of radius $\frac{1}{n}$, $f_n(x) = 0$ if $\|x-y\| \ge \frac{1}{n}$ for all $y \in B_r$ and thus f_n has support in the ball of radius $r + \frac{1}{n}$. This establishes Assertion 1.

Let $\varepsilon > 0$ be given and let $|\alpha| \le k$. Since $\partial_x^\alpha f$ is uniformly continuous, we may choose N so that if $n \ge N$ and if $\|y\| \le \frac{1}{n}$, then $|\partial_x^\alpha f(x) - \partial_x^\alpha f(x-y)| < \varepsilon$. As $\int_{\mathbb{R}^m} g_n = 1$,

$$\partial_x^\alpha f(x) - \partial_x^\alpha f_n(x) = \int_{\mathbb{R}^m} g_n(y) \partial_x^\alpha f(x) dy - \int_{\mathbb{R}^m} g_n(y) \partial_x^\alpha f(x-y) dy.$$

Because g_n has support in $B_{\frac{1}{n}}$ and g_n is non-negative, we complete the proof by computing

$$|\partial_x^\alpha f(x) - \partial_x^\alpha f_n(x)| \le \int_{\mathbb{R}^m} g_n(y) |\partial_x^\alpha f(x) - \partial_x^\alpha f(x-y)| dy$$

$$= \int_{\|y\| \le \frac{1}{n}, y \in \mathbb{R}^m} g_n(y) |\partial_x^\alpha f(x) - \partial_x^\alpha f(x-y)| dy$$

$$\le \varepsilon \int_{\mathbb{R}^m} g_n(y) dy = \varepsilon. \qquad \square$$

17.2   SOBOLEV SPACES AND THE RELLICH LEMMA

In Section 17.2.1, we define the Sobolev spaces $H_s(\mathbb{R}^m)$; these are the completion of the Schwarz space $\mathcal{S}$ with respect to the Sobolev norms $\|\cdot\|_{s,L^2}$. In Section 17.2.2, we establish the Sobolev Lemma which shows that functions which have sufficiently many L^2 derivatives in fact are differentiable in the ordinary sense. In Section 17.2.3, we establish the Rellich Lemma which shows the inclusion of $H_{s,L^2}(\mathbb{R}^m)$ in $H_{t,L^2}(\mathbb{R}^m)$ is a compact injection for $s > t$ if we uniformly bound the supports; this technical requirement on the supports does not play an essential role in our development since we will be interested subsequently in the case of compact manifolds.

17.2.1 SOBOLEV SPACES. If $f \in \mathcal{S}$, if $s \in \mathbb{R}$, and if k is a non-negative integer, set

$$\|f\|_{s,L^2} := \|(1+\|\xi\|^2)^{\frac{1}{2}s} \hat{f}\|_{L^2}, \quad \|f\|_{C^k} := \sum_{|\alpha| \le k} \sup_{x \in \mathbb{R}^m} |\partial_x^\alpha f(x)|,$$

$$\tilde{\|} f \tilde{\|}_{k,L^2}^2 := \sum_{|\alpha| \le k} \|\partial_x^\alpha f\|_{L^2}^2, \qquad \nu_s := (1+\|\xi\|^2)^{\frac{s}{2}} d\xi.$$

The *Sobolev space* $H_s(\mathbb{R}^m)$ is the completion of $\mathcal{S}$ with respect to the *Sobolev norm* $\|\cdot\|_{s,L^2}$. Since $\mathcal{S}$ is dense in $L^2(\nu_s)$, the Fourier transform identifies $H_s(\mathbb{R}^m)$ with $L^2(\nu_s)$.

Lemma 17.8 If k is a non-negative integer, then $\tilde{\|}_{k,L^2}$ and $\|_{k,L^2}$ are equivalent norms.

Proof. By Lemmas 17.2 and 17.4, $\|\partial_x^\alpha f\|_{L^2}^2 = \|\widehat{\partial_x^\alpha f}\|_{L^2}^2 = \|\xi^\alpha \hat{f}\|_{L^2}^2$. Thus,

$$\sum_{|\alpha|\leq k} \|\partial_x^\alpha f\|_{L^2}^2 = \int_{\mathbb{R}^m} \sum_{|\alpha|\leq k} \xi^{2\alpha} |f(\xi)|^2 d\xi.$$

Since there exist positive constants ε_i so $\varepsilon_1 (1 + \|\xi\|^2)^k \leq \sum_{|\alpha|\leq k} \xi^{2\alpha} \leq \varepsilon_2 (1 + \|\xi\|^2)^k$,

$$\varepsilon_1 \|f\|_{k,L^2}^2 \leq \sum_{|\alpha|\leq k} \|\partial_x^\alpha f\|_{L^2}^2 \leq \varepsilon_2 \|f\|_{k,L^2}^2. \qquad \square$$

Lemma 17.8 shows that $\|f\|_{k,L^2}$ measures the L^2 derivatives of f up to order k. The Laplacian is defined by setting $\Delta := -\partial_{x^1}^2 - \cdots - \partial_{x^m}^2$. Because the Fourier transform of $(1 + \Delta)^k f(\xi)$ is $(1 + \|\xi\|^2)^k \hat{f}(\xi)$,

$$\|f\|_{2k,L^2} = \|(1 + \|\xi\|^2)^k \hat{f}\|_{L^2} = \|(1 + \Delta)^k f\|_{L^2}. \qquad (17.2.\text{a})$$

Equation (17.2.a) is closely related to Gårding's inequality as we shall see in Section 17.4.3.

Lemma 17.9

1. ∂_x^α defines a continuous map from $H_s(\mathbb{R}^m)$ to $H_{s-|\alpha|}(\mathbb{R}^m)$.

2. The map $f \otimes g \to (f, g)_{L^2}$ extends to a perfect pairing $H_s(\mathbb{R}^m) \otimes H_{-s}(\mathbb{R}^m) \to \mathbb{C}$ that exhibits each space as the dual of the other, i.e., if $f \in H_{s,L^2}$, then

$$\|f\|_{s,L^2} = \sup_{0\neq g\in\mathcal{S}} \frac{|(f,g)_{L^2}|}{\|g\|_{-s,L^2}}.$$

3. If $\vartheta \in \mathcal{S}$, the map $f \to \vartheta f$ for $f \in \mathcal{S}$ extends to a continuous linear map m_ϑ from $H_s(\mathbb{R}^m)$ to $H_s(\mathbb{R}^m)$.

Proof. As $|\xi_k| \leq \|\xi\| \leq (1 + \|\xi\|^2)^{\frac{1}{2}}$, we prove Assertion 1 by using Lemma 17.2 to see

$$\begin{aligned}
\|\partial_x^\alpha f\|_{s-|\alpha|}^2 &= \|(1 + \|\xi\|^2)^{\frac{1}{2}(s-|\alpha|)} \widehat{\partial_x^\alpha f}\|_{L^2}^2 = \|(1 + \|\xi\|^2)^{\frac{1}{2}(s-|\alpha|)} \xi^\alpha f\|_{L^2}^2 \\
&\leq \|(1 + \|\xi\|^2)^{\frac{1}{2}(s-|\alpha|)} (1 + \|\xi\|^2)^{\frac{1}{2}|\alpha|} f\|_{L^2}^2 = \|(1 + \|\xi\|^2)^{\frac{1}{2}s} f\|_{L^2}^2 = \|f\|_{s,L^2}.
\end{aligned}$$

Let $f \in \mathcal{S}$ and $g \in \mathcal{S}$. By Lemma 17.4 and the Cauchy–Schwarz–Bunyakovsky inequality,

$$\begin{aligned}
|(f,g)|_{L^2} &= |(\hat{f}, \hat{g})_{L^2}| = \left| \int_{\mathbb{R}^m} \hat{f}(\xi) \bar{\hat{g}}(\xi) d\xi \right| \leq \|(1 + \|\xi\|^2)^{\frac{1}{2}s} \hat{f}\|_{L^2} \|(1 + \|\xi\|^2)^{-\frac{s}{2}} \hat{g}\|_{L^2} \\
&= \|f\|_{s,L^2} \|g\|_{-s,L^2}.
\end{aligned}$$

This shows that

$$\sup_{0\neq g\in\mathcal{S}} \frac{|(f,g)_{L^2}|}{\|g\|_{-s,L^2}} \leq \|f\|_{s,L^2}. \qquad (17.2.\text{b})$$

Let $f \in \mathcal{S}$. To show that equality holds in Equation (17.2.b), we use the Fourier inversion formula to choose g so $\hat{g} = (1 + \|\xi\|^2)^s \hat{f}$. We then have

$$\|g\|_{-s,L^2} = \|\hat{g}(1 + \|\xi\|^2)^{-\frac{1}{2}s}\|_{L^2} = \|\hat{f}(1 + \|\xi\|^2)^{\frac{1}{2}s}\| = \|f\|_{s,L^2},$$

$$(f, g)_{L^2} = (\hat{f}, \hat{g})_{L^2} = (\hat{f}, (1 + \|\xi\|^2)^s \hat{f})_{L^2} = \|f\|_{s,L^2}^2 = \|f\|_{s,L^2} \|g\|_{-s,L^2}.$$

This establishes Assertion 2 if $f \in \mathcal{S}$. Since $H_s(\mathbb{R}^m)$ is the completion of $\mathcal{S}$ with respect to $\|\cdot\|_{s,L^2}$, Assertion 2 follows in complete generality.

We now establish Assertion 3. Let $f \in \mathcal{S}$. The argument given to prove Assertion 2 shows that we may choose h so $\|h\|_{-s,L^2} = 1$ and so $(f\vartheta, h)_{L^2} = \|f\vartheta\|_{s,L^2}$. We use Lemma 17.6 to see that

$$(f\vartheta, h)_{L^2} = \int_{\mathbb{R}^m} \widehat{f\vartheta}(\xi)\bar{\hat{h}}(\xi)d\xi = \int_{\mathbb{R}^m} (\hat{f} \star \hat{\vartheta})\bar{\hat{h}}(\xi)d\xi$$

$$= (2\pi)^{-\frac{1}{2}m} \int_{\mathbb{R}^m} \int_{\mathbb{R}^m} \hat{f}(\eta)\hat{\vartheta}(\xi - \eta)\bar{\hat{h}}(\xi)d\eta d\xi. \tag{17.2.c}$$

Let $k > \frac{1}{2}|s|$. We use Lemma 17.1 and Peetre's inequality (Lemma 16.27) to estimate:

$$|\hat{\vartheta}(\xi - \eta)| (1 + \|\eta\|^2)^{-\frac{1}{2}s} (1 + \|\xi\|^2)^{\frac{1}{2}s}$$
$$\le C(\vartheta)(1 + \|\xi - \eta\|^2)^{-k} (1 + \|\eta\|^2)^{-\frac{1}{2}s} (1 + \|\xi\|^2)^{\frac{1}{2}s} \tag{17.2.d}$$
$$\le C_1(s, \vartheta)(1 + \|\xi - \eta\|^2)^{-k} (1 + \|\xi - \eta\|^2)^{\frac{1}{2}|s|} \le C_1(s, \vartheta).$$

We complete the proof by using the Cauchy–Schwarz–Bunyakovsky inequality and Equations (17.2.c)–(17.2.d) to estimate

$$\|f\vartheta\|_{s,L^2} = (f\vartheta, h)_{L^2} \le C_2 \int_{\mathbb{R}^{2m}} |\hat{f}(\eta)| |\hat{\vartheta}(\xi - \eta)| |\hat{h}(\xi)|d\eta d\xi$$

$$= C_2 \int_{\mathbb{R}^{2m}} \left\{|\hat{f}(\eta)|(1 + \|\eta\|^2)^{\frac{1}{2}s}\right\} \cdot \left\{|\hat{h}(\xi)|(1 + \|\xi\|^2)^{-\frac{1}{2}s}\right\}$$

$$\cdot \left\{|\hat{\vartheta}(\xi - \eta)|(1 + \|\eta\|^2)^{-\frac{1}{2}s}(1 + \|\xi\|)^{\frac{1}{2}s}\right\} d\eta d\xi$$

$$\le C_3(s, \vartheta) \int_{\mathbb{R}^{2m}} \left\{|\hat{f}(\eta)|(1 + \|\eta\|^2)^{\frac{1}{2}s}\right\} \cdot \left\{|\hat{h}(\xi)|(1 + \|\xi\|^2)^{-\frac{1}{2}s}\right\} d\eta d\xi$$

$$\le C_3(s, \vartheta)\|f\|_{s,L^2} \|h\|_{-s,L^2} = C_4(s, \vartheta)\|f\|_{s,L^2}. \qquad \square$$

17.2.2 THE SOBOLEV LEMMA FOR $\mathbb{R}^m$ [68].

Lemma 17.10   Let k be a positive integer and let $s > k + \frac{1}{2}m$.

1. $\|f\|_{C^k} \le C(k, s, m)\|f\|_{s,L^2}$.
2. The inclusion of $\mathcal{S}$ into $C^k(\mathbb{R}^m)$ extends to a continuous injective map from $H_s(\mathbb{R}^m)$ to $C^k(\mathbb{R}^m)$ which permits us to regard $H_s(\mathbb{R}^m)$ as a subspace of $C^k(\mathbb{R}^m)$.

Proof. Suppose first $k = 0$ and $s > \frac{m}{2}$. Then $(1 + \|\xi\|^2)^{-\frac{1}{2}s} \in L^2$. We use the Fourier inversion formula given in Lemma 17.3 to compute:

$$
\begin{aligned}
|f(x)| &= (2\pi)^{-\frac{1}{2}m} \left| \int_{\mathbb{R}^m} e^{\sqrt{-1}\, x \cdot \xi} \hat{f}(x) dx \right| \\
&\leq (2\pi)^{-\frac{1}{2}m} \|(1 + \|\xi\|^2)^{\frac{1}{2}s} \hat{f}\|_{L^2} \cdot \|(1 + \|\xi\|^2)^{-\frac{s}{2}}\|_{L^2} \\
&\leq C(0, s, m) \|f\|_{s, L^2} \,.
\end{aligned}
$$

More generally, we can use Lemma 17.9 to establish Assertion 1 by estimating

$$
|\partial_x^\alpha f(x)| \leq C(0, s - k, m)\|\partial_x^\alpha f\|_{s-k} \leq C(k, s, m)\|f\|_{s, L^2} \,.
$$

The Sobolev space $H_s(\mathbb{R}^m)$ is the completion of $\mathcal{S}$ with respect to the norm $\|\cdot\|_{s, L^2}$. The uniform limit of C^k functions is again C^k. Thus, by Assertion 1, the natural inclusion of $\mathcal{S}$ into $C^k(\mathbb{R}^m)$ extends to a continuous linear map ι from $H_s(\mathbb{R}^m)$ to $C^k(\mathbb{R}^m)$. Let $f \in H_s(\mathbb{R}^m)$. Choose elements $\phi_n \in \mathcal{S}$ so that $\phi_n \to f$ in $\|\cdot\|_{s, L^2}$ and thus $\phi_n \to \iota(f)$ in $C^k(\mathbb{R}^m)$. If $\iota(f) = 0$, then $\phi_n \to 0$ uniformly. Let $g \in \mathcal{S} \subset L^2$. By Lebesgue dominated convergence,

$$
(f, g)_{L^2} = \lim_{n \to \infty} (\phi_n, g)_{L^2} = 0 \,.
$$

Consequently, $f = 0$ by Lemma 17.9. This shows ι is 1-1 so we can regard $H_s(\mathbb{R}^m)$ as a subspace of $C^k(\mathbb{R}^m)$.

17.2.3 THE RELLICH LEMMA FOR $\mathbb{R}^m$ [62].

Lemma 17.11   If $\vartheta \in \mathcal{S}$, let $m_\vartheta(f) = \vartheta f$. Let $t < s$.

1. The identity map of $\mathcal{S}$ induces a continuous injective linear map from $H_s(\mathbb{R}^m)$ to $H_t(\mathbb{R}^m)$ so we can regard $H_s(\mathbb{R}^m)$ as a linear subspace of $H_t(\mathbb{R}^m)$.

2. m_ϑ is a compact linear map from $H_s(\mathbb{R}^m)$ to $H_t(\mathbb{R}^m)$.

3. Let K be a compact subset of $\mathbb{R}^m$ and let $f_n \in C_0^\infty(K) \subset \mathcal{S}$ satisfy $\|f_n\|_{s, L^2} \leq 1$. Then there exists a subsequence f_{n_k} which converges in H_t.

Proof. If $t < s$, then $(1 + \|\xi\|^2)^t \leq (1 + \|\xi\|^2)^s$ so $\|\cdot\|_{t, L^2} \leq \|\cdot\|_{s, L^2}$. Thus, the identity map of $\mathcal{S}$ induces a continuous map $\iota_{s,t}$ from $H_s(\mathbb{R}^m)$ to $H_t(\mathbb{R}^m)$. Let

$$
v_s := (1 + \|\xi\|^2)^{\frac{1}{2}s} \,.
$$

The Fourier transform identifies $H_s(\mathbb{R}^m) = L^2(v_s)$ and $\iota_{s,t}$ is induced from the identity map from $L^2(v_s) \to L^2(v_t)$. A measurable function which vanishes almost everywhere with respect to the measure v_s vanishes almost everywhere with respect to the measure v_t and, consequently, $\iota_{s,t}$ is injective; we can regard $H_s(\mathbb{R}^m)$ as a linear subspace of $H_t(\mathbb{R}^m)$.

By Lemma 17.9, $m_\vartheta \in \mathrm{Hom}(H_s(\mathbb{R}^m), H_s(\mathbb{R}^m))$. As the identity map of $\mathcal{S}$ induces a continuous linear map from $H_s(\mathbb{R}^m)$ to $H_t(\mathbb{R}^m)$, $m_\vartheta \in \mathrm{Hom}(H_s(\mathbb{R}^m), H_t(\mathbb{R}^m))$. Let f_n be a bounded sequence of elements of $H_s(\mathbb{R}^m)$. We wish to show a subsequence of the functions $m_\vartheta f_n$ converges in $H_t(\mathbb{R}^m)$. Since $H_s(\mathbb{R}^m)$ is the completion of $\mathcal{S}$ in the $\|\cdot\|_{s,L^2}$ norm, we may assume without loss of generality that $f_n \in \mathcal{S}$. By rescaling the bound, we suppose $\|f_n\| \leq 1$. By Lemma 17.6, $\widehat{m_\vartheta f} = \hat{f} \star \hat{\vartheta}$. The integrals in question converge uniformly so we can differentiate under the integral sign and apply the Cauchy–Schwarz–Bunyakovsky inequality to see

$$\left|\{\partial_\xi^\alpha(\widehat{m_\vartheta f_n})\}(\xi)\right| = (2\pi)^{-\frac{1}{2}m}\left|\partial_\xi^\alpha \int_{\mathbb{R}^m} \hat{f}_n(\eta)\hat{\vartheta}(\xi - \eta)d\eta\right|$$

$$= (2\pi)^{-\frac{1}{2}m}\left|\int_{\mathbb{R}^m} \hat{f}_n(\eta)(\partial_\xi^\alpha \hat{\vartheta})(\xi - \eta)d\eta\right| \leq C(\xi, \alpha, \vartheta)\|f_n\|_{s,L^2}$$

where

$$C(\xi, \alpha, \vartheta)^2 := (2\pi)^{-\frac{1}{2}m}\int_{\mathbb{R}^m} |(\partial_\xi^\alpha \hat{\vartheta})(\xi - \eta)|^2(1 + \|\eta\|^2)^{-s}d\eta.$$

Choose k so $|s| - k < -\frac{1}{2}m$. By Lemma 17.1 (3) and Peetre's inequality,

$$C(\xi, \alpha, \vartheta)^2 \leq C_1(s, k, \alpha, \vartheta)\int_{\mathbb{R}^m}(1 + \|\xi - \eta\|^2)^{-k}(1 + \|\eta\|^2)^{-s}d\eta$$

$$\leq C_2(s, k, \alpha, \vartheta)(1 + \|\xi\|^2)^{|s|}\int_{\mathbb{R}^m}(1 + \|\eta\|^2)^{|s|-k}d\eta$$

$$\leq C_3(s, k, \alpha, \vartheta)(1 + \|\xi\|^2)^{|s|}.$$

This shows that all the derivatives of the functions $\widehat{m_\vartheta f_n}$ are uniformly bounded on compact ξ subsets. The Arzelà–Ascoli Theorem (see Theorem 16.15) shows that there exists a subsequence f_n so that $\widehat{m_\vartheta f_n}$ converges uniformly on compact ξ subsets. To simplify the notation, we replace the original sequence by this subsequence. Let $\varepsilon > 0$ be given. We have

$$\|m_\vartheta f_i - m_\vartheta f_j\|_{t,L^2}^2 = \int_{\mathbb{R}^m}(1 + \|\xi\|^2)^t |\widehat{m_\vartheta f_i} - \widehat{m_\vartheta f_j}|^2(\xi)d\xi. \qquad (17.2.e)$$

Choose r so that $(1 + r^2)^{t-s} < \varepsilon$. We decompose the integral of Equation (17.2.e) into two pieces $\boxed{1}$ where $\|\xi\| \geq r$ and $\boxed{2}$ where $\|\xi\| \leq r$. We estimate

$$\boxed{1} := \int_{\|\xi\| \geq r}(1 + \|\xi\|^2)^t |\widehat{m_\vartheta f_i} - \widehat{m_\vartheta f_j}|^2(\xi)d\xi$$

$$= \int_{\|\xi\| \geq r}(1 + \|\xi\|^2)^{t-s}(1 + \|\xi\|^2)^s |\widehat{m_\vartheta f_i} - \widehat{m_\vartheta f_j}|^2(\xi)d\xi$$

$$\leq (1 + r^2)^{t-s}\int_{\|\xi\| \geq r}(1 + \|\xi\|^2)^s |\widehat{m_\vartheta f_i} - \widehat{m_\vartheta f_j}|^2(\xi)d\xi$$

$$\leq \varepsilon\|\widehat{m_\vartheta f_i} - \widehat{m_\vartheta f_j}\|_{s,L^2}^2 \leq \varepsilon\|\|m_\vartheta\|\|\, \|f_i - f_j\|_{s,L^2}^2.$$

By assumption, $\|f_i\|_{s,L^2} \leq 1$ and, consequently, $\|f_i - f_j\|^2_{s,L^2} \leq 4$. Thus,

$$\boxed{1} = \int_{\|\xi\| \geq r} (1 + \|\xi\|^2)^t |\widehat{m_\vartheta f_i} - \widehat{m_\vartheta f_j}|^2(\xi) d\xi \leq 4\varepsilon \|\|m_\vartheta\|\|. \qquad (17.2.\text{f})$$

Because the sequence $\{\widehat{m_\vartheta f_i}\}$ converges uniformly on compact ξ subsets,

$$\boxed{2} := \int_{\|\xi\| \leq r} (1 + \|\xi\|^2)^t |\widehat{m_\vartheta f_i} - \widehat{m_\vartheta f_j}| d\xi < \varepsilon \quad \text{if} \quad i, j > N(\varepsilon). \qquad (17.2.\text{g})$$

Assertion 2 now follows from Equations (17.2.f) and (17.2.g).

To prove Assertion (3), we suppose that $\{f_n\}$ is a sequence of elements in $\mathcal{S}$ such that $\|f_n\|_{s,L^2} \leq 1$ and such that the support of f_n is contained in a fixed compact set K. Choose $\vartheta \in C_0^\infty(\mathbb{R}^m)$ so that $\vartheta \equiv 1$ on K (see Section 1.6 of Book I); ϑ is called a *plateau function*. Then $m_\vartheta(f_n) = f_n$. Assertion 3 now follows from Assertion 2. $\qquad \square$

17.3 NORMS ON SPACES OF SECTIONS TO A BUNDLE

Let (V, h, ∇) be a geometric vector bundle over a compact Riemannian manifold (M, g) and let k be a non-negative integer. Let $C^k(V)$ be the space of C^k sections to V. In Section 17.3.1, we will discuss a norm $\|\cdot\|_{k,\infty}$ which makes $C^k(V)$ into a Banach space. In Section 17.3.2, we use convolution smoothing to prove $C^\infty(V)$ is dense in $C^k(V)$ for any k and to show $C^k(V)$ is separable. In Section 17.3.3, we define a norm $\|\cdot\|_{k,L^2}$ on $C^\infty(V)$ to measure L^2 derivatives. The completion of $C^\infty(V)$ with respect to $\|\cdot\|_{k,L^2}$ is the Sobolev space $H_k(V)$. We note that the Sobolev spaces $H_s(V)$ can be defined for any $s \in \mathbb{R}$; as we shall not need these spaces, we omit any further discussion of this generalization. In Section 17.3.4, we examine integration by parts and interpolation. In Section 17.3.5, we introduce operators of Laplace type and present the Bochner formalism.

17.3.1 THE SUP NORM ON $C^k(V)$.
We generalize Example 16.25. If $\phi \in C(V)$, let $\|\phi(x)\|_h := h(\phi(x), \phi(x))^{\frac{1}{2}}$ and $\|\phi\|_{h,\infty} := \sup_{x \in M} \|\phi(x)\|_h$. If $\phi \in C^k(V)$, set

$$\|\phi\|_{g,h,\nabla,C^k} := \sum_{j=0}^k \|\nabla^j \phi\|_{g,h,\infty}.$$

Lemma 17.12 Let (V, h, ∇) be a geometric vector bundle over a compact Riemannian manifold (M, g).

1. Let $\Psi : C^\infty(V) \to C^\infty(W)$ be a kth-order partial differential operator. Then
$$\|\Psi\phi\|_{g_W,h_W,\nabla_W,C^j} \leq C \|\phi\|_{g_V,h_V,\nabla_V,C^{j+k}}.$$
2. $C^k(V)$ is complete with respect to the norm $\|\cdot\|_{g,h,\nabla,C^k}$.

3. Let $\{g_i, h_i, \nabla_i\}$ be suitable structures on (M, V) for $i = 1, 2$. Then

$$\|\cdot\|_{g_2,h_2,\nabla_2,C^k} \leq C \|\cdot\|_{g_1,h_1,\nabla_1,C^k} .$$

4. The Banach space structure on $C^k(V)$ is independent of the choice of $\{g, h, \nabla\}$.

Proof. We adopt the following notation for the remainder of the proof. Let $P \in M$. Choose smooth local frames $\vec{s} = (s^1, \ldots, s^\ell)$ and $\vec{t} = (t^1, \ldots, t^n)$ for V and W, respectively, over a compact neighborhood K_P of P. By applying the Gram–Schmidt process we may assume that $\vec{s}$ and $\vec{t}$ are orthonormal frames. Expand $\phi \in C(V)$ in the form $\phi = \phi_\nu s^\nu$ where we adopt the *Einstein convention* and sum over repeated indices; $\|\phi\|_h^2 = h(\phi, \phi) = \sum_\nu |\phi_\nu|^2$. As always, C will be a generic constant which can increase at each step in an estimation. To prove Assertion 1, we first suppose that $j = 0$ and $k = 0$ so $\Psi \in \mathrm{Hom}(V, W)$. Then $\Psi(\phi_\nu s^\nu) = \Psi_\mu^\nu \phi_\nu t^\mu$ where $\Psi_\mu^\nu := h_W(\Psi s^\nu, t^\mu)$. Since K_P is compact, the constant $C_K := \max_{x \in K} |\Psi_\mu^\nu(x)| < \infty$ is well defined. By the Cauchy–Schwarz–Bunyakovsky inequality,

$$\|\Psi\phi(x)\|_{h_W}^2 = \sum_\mu \left| \sum_\nu \Psi_\nu^\mu \phi_\nu(x) \right|^2 \leq \sum_\mu \left\{ \sum_\nu |\Psi_\nu^\mu|^2 \right\} \cdot \left\{ \sum_\nu |\phi_\nu(x)|^2 \right\}$$

$$\leq \dim(V) \dim(W) C_K^2 \|\phi(x)\|_{h_V}^2 .$$

Since M is compact, we can cover M by a finite number of such charts and take the maximum of the constants to obtain $\|\Psi\phi(x)\|_{h_W}^2 \leq C \|\phi(x)\|_{h_V}^2$. Consequently,

$$\|\Psi\phi\|_{h_W,\infty}^2 = \sup_{x \in M} \|\Psi\phi(x)\|_{h_W}^2 \leq C \sup_{x \in M} \|\phi(x)\|_{h_V}^2 = C \|\phi\|_{h_V,\infty}^2 .$$

More generally, if $j = 0$ but $k > 0$, we may express $\Psi = \sum_{i=0}^{k} \Psi_i \circ \nabla^i$ where Ψ_i belongs to $\mathrm{Hom}(\otimes^k T^*M \otimes V, W)$. We may then estimate

$$\|\Psi\phi\|_{h_W,\infty} \leq \sum_{i=0}^{k} C \|\nabla_V^i \phi\|_{g,h_V,\infty} \leq C \|\phi\|_{g,h_V,\nabla_V,C^k} .$$

If $j > 0$, we replace Ψ by $\nabla_V^j \Psi$ to obtain a pointwise estimate yielding Assertion 1:

$$\|\Psi\phi(x)\|_{g,h_W,\nabla_W,C^j} \leq \|\phi(x)\|_{g,h_V,\nabla_V,C^{k+j}} . \tag{17.3.a}$$

Since the uniform limit of continuous functions is continuous, $(C(V), \|\cdot\|_{h,\infty})$ is complete. Let $\vec{x} = (x^1, \ldots, x^m)$ be a system of coordinates on K_P. By Assertion 1, we can estimate $\|\partial_x^\alpha \phi(x)\|_h \leq C \|\phi\|_{g,h,\nabla_V,C^k}$ for $|\alpha| \leq k$. Since the uniform limit of C^k functions is C^k, we conclude $(C^k(V), \|\cdot\|_{g,h,\nabla,C^k})$ is complete. This proves Assertion 2.

Let (g_i, h_i, ∇_i) be suitable structures on (M, V) for $i = 1, 2$. Suppose first that $k = 0$ so $\|\phi(x)\|_{h_1}^2 = \sum_\nu |\phi_\nu(x)|^2$. Let $h_2^{\mu\nu} = h(s^\mu, s^\nu)$. Find C so that $|h^{\mu\nu}(x)| \leq C$ for $1 \leq \mu, \nu \leq \ell$ and $x \in K_P$. By Equation (16.3.a),

$$\|\phi\|_{h_2}^2 = \sum_{\mu,\nu} h^{\mu\nu} \phi_\mu \phi_\nu \leq C \sum_{\mu\nu} |\phi_\mu \phi_\nu| \leq C \sum_\mu |\phi_\mu|^2 \leq C \|\phi\|_{h_1}^2.$$

By covering M by a finite collection of the K_P we obtain

$$\|\phi\|_{g_2, h_2, \infty} \leq C \|\phi\|_{g_1, h_1, \infty}.$$

Replacing ϕ by $\nabla^k \phi$ and the bundle V by $\otimes^k T^* M \otimes V$ and applying Assertion 1 yields

$$
\begin{aligned}
\|\phi\|_{g_2, h_2, \nabla_2, C^k} &= \sum_{j \leq k} \|\nabla_2^j \phi\|_{g_2, h_2, \infty} \leq C \sum_{j \leq k} \|\nabla_2^j \phi\|_{g_1, h_1, \infty} \\
&\leq C \sum_{j \leq k} \|\nabla_1^j \phi\|_{g_1, h_1, \infty} = C \|\phi\|_{g_1, h_1, \nabla_1, C^k}.
\end{aligned}
$$

This proves Assertion 3; Assertion 4 now follows. □

Since the Banach space structure on $C^k(V)$ defined by the norms $\|\cdot\|_{g, h, \nabla, C^k}$ is independent of (g, h, ∇), to avoid notational complexity, we simply denote any of these norms by $\|\cdot\|_{C^k}$ henceforth. We give $C^\infty(V)$ the *Fréchet space* topology given in Definition 16.23 that is defined by the sequence of increasing norms $\|\cdot\|_{C^k}$; this topology is metrizable by Lemma 16.24.

17.3.2 APPROXIMATIONS.

Lemma 17.13   Let (V, h, ∇) be a geometric vector bundle over a compact Riemannian manifold (M, g).

1. $C^\infty(V)$ is dense in $C^k(V)$ for any k.
2. $C^k(V)$ is a separable Banach space.

Proof. Let $B_{3r(P)}(P)$ be the geodesic ball of radius $3r(P)$ about P in M where $r(P) > 0$ is chosen so that geodesic coordinates are well defined on $B_{3r(P)}(P)$. By shrinking $r(P)$ if necessary, we can assume that V admits a local orthonormal frame on $B_{3r(P)}(P)$. Choose a partition of unity ψ_i subordinate to the open cover $\{B_{r(P)}(P)\}_{P \in M}$. Let $\phi \in C^k(V)$. Express $\psi_i \phi = \phi_{i,\mu} s_i^\mu$ for $\phi_{i,\mu} \in C_0^k(B_{r(P_i)}(P_i))$. By Lemma 17.7, there exist smooth functions $\phi_{n,i,\mu} \in C_0^\infty(B_{2r(P_i)}(P_i))$ so that the sequence $\{\partial_x^\alpha \phi_{n,i,\mu}\}$ converges uniformly to $\partial_x^\alpha \phi_{i,\mu}$ on $B_{2r(P_i)}(P_i)$ for $|\alpha| \leq k$ as $n \to \infty$. Let $\phi_{n,i} := \phi_{n,i,\mu} s^\mu$; $\phi_{n,i}$ a smooth section to V with support in $B_{r(P_i)}(P_i)$. The sequence $\{\phi_{n,i}\}$ converges uniformly to $\psi_i \phi$ in the C^k topology as $n \to \infty$. Sum over i to complete the proof of Assertion 1.

We continue the discussion to prove Assertion 2. Fix i and embed $B_{3r(P_i)}(P_i)$ in the torus $\mathbb{T}^m := S^1 \times \cdots \times S^1$. By Assertion 1, we can uniformly approximate $\phi_{i,\mu}$ by a smooth function $\tilde{\phi}_{i,\mu}$ with compact support in $B_{2r(P_i)}(P_i)$. The argument of Example 16.33 extends to show we can approximate $\tilde{\phi}_{i,\mu}$ in the C^k topology by a trigonometric polynomial on $\mathbb{T}^m$. Let $\mathcal{P}$ be the countable set of trigonometric polynomials with rational coefficients. We can uniformly approximate $\tilde{\phi}_{i,\mu}$ in the C^k topology by elements of $\mathcal{P}$. Let $\vartheta_i \in C_0^\infty(B_{3r(P_i)}(P_i))$ be *plateau functions* which are identically 1 on $B_{2r(P_i)}(P_i)$. Since $\vartheta_i \tilde{\phi}_{i,\mu} = \tilde{\phi}_{i,\mu}$, we can uniformly approximate $\tilde{\phi}_{i,\mu}$ by elements of $\vartheta_i \mathcal{P}$ on $B_{3r(P_i)}(P_i)$ in the C^k topology. Thus, we can uniformly approximate $\phi_{i,\mu}$ by elements of $\vartheta_i \mathcal{P}$ in the C^k topology. We then pass back to the global setting to complete the proof of Assertion 2. □

17.3.3 MEASURING L^2 DERIVATIVES. Let $L^2(V, g, h)$ be the Hilbert space completion of $C^\infty(V)$ with respect to the L^2 inner product $(\phi, \psi)_{g,h} = \int_M (\phi, \psi)_h \, |\mathrm{dvol}|(g)$. More generally, set

$$(\phi, \psi)_{g,h,\nabla,k,L^2} = \int_M \sum_{j \le k} (\nabla^j \phi, \nabla^j \psi)_{g,h} \, |\mathrm{dvol}|(g).$$

Let $H_{g,h,\nabla,k}(V)$ be the Hilbert space completion of $C^\infty(V)$ with respect to this inner product. We have $C^k(V) \subset H_{g,h,\nabla,k}(V)$.

Lemma 17.14   Let V be a vector bundle over a compact manifold M. Let (g_i, h_i, ∇_i) for $i = 1, 2$ be suitable structures on (M, V).

1. $\|\cdot\|_{g,h,\nabla,k,L^2}^2 \le \|\cdot\|_{g,h,\nabla,C^k}^2 \cdot \mathrm{Vol}(M)$. Thus, the inclusion map defines an injective linear map $\iota_{C^k, H_k} : C^k(V) \to H_{g,h,\nabla,k}(V)$ with dense range so $H_{g,h,\nabla,k}$ is separable.

2. $\|\cdot\|_{g_2,h_2,\nabla_2,k,L^2} \le C\|\cdot\|_{g_1,h_1,\nabla_1,k,L^2}$. Thus, the underlying Banach structure on $H_{g,h,\nabla,k}$ is independent of the choice of $\{g, h, \nabla\}$.

3. Let $P : C^\infty(V) \to C^\infty(W)$ be a kth-order partial differential operator. Then we have that $\|P\phi\|_{g,h_W,\nabla_W,j,L^2} \le C\|\phi\|_{g,h_V,\nabla_V,k+j,L^2}$ so P extends to a continuous linear map from $H_{g,h_V,\nabla_V,j+k,L^2}(V)$ to $H_{g,h_W,\nabla_W,j,L^2}(V)$ for any k.

Proof. The estimate of Assertion 1 is immediate from the definition. Thus, ι_{C^k, H_k} is a continuous linear map; it is immediate from the definition that the range is dense. By Lemma 17.13, $C^k(V)$ is separable; thus, $H_{g,h,\nabla,k}(V)$ is separable. Suppose $\phi \in C^k(V)$ and $\iota_{C^k, H_k}(\phi) = 0$. Since $\|\cdot\|_{g,h,\nabla,0,L^2} \le \|\cdot\|_{g,h,\nabla,k,L^2}$, we have $\phi = 0$ in L^2. This implies $\phi = 0$ almost everywhere and, as ϕ is continuous, $\phi = 0$. This implies ι_{C^k, H_k} is injective.

The proof of Lemma 17.12 gives rise to a pointwise estimate of the form

$$\sum_{j \le k} \|\nabla_2^j \phi(x)\|_{g_2,h_2}^2 \le C \sum_{j \le k} \|\nabla_1^j \phi(x)\|_{g_1,h_1}^2.$$

Since $|\mathrm{dvol}|(g_2) \le C|\mathrm{dvol}|(g_1)$, we can integrate this estimate to establish the estimate of Assertion 2 and complete thereby the proof of Assertion 2. We square the pointwise estimate of Equation (17.3.a) and integrate to obtain Assertion 3. □

We set $H_k(V) := H_{g,h,\nabla,k,L^2}$ and $\|\cdot\|_{k,L^2} := \|\cdot\|_{g,h,\nabla,k,L^2}$ for any (g, h, ∇) since, by Lemma 17.14, the underlying Banach space structure is independent of (g, h, ∇). These are the *Sobolev spaces*; they will play an important role in our treatment of the spectral theory of self-adjoint operators of Laplace type subsequently.

17.3.4 INTEGRATION BY PARTS AND INTERPOLATION.

Definition 17.15   Let $\phi \in C^\infty(V)$. We work locally and choose a local orthonormal frame for V over a coordinate chart $\mathcal{O}$. This permits us to regard ϕ as a map from $\mathcal{O}$ to $\mathbb{C}^\ell$ and define $\partial_{x^i}\phi$. Expand $\nabla_{\partial_{x^i}}\phi = \partial_{x^i}\phi + \omega_i\phi$ where $\omega = \omega_i\,dx^i$ is the connection 1-form of ∇. Let $\Gamma_{ij}{}^k$ be the Christoffel symbols of the Levi–Civita connection (see Lemma 16.2). We may expand $\nabla\phi = \phi_{;i}\,dx^i$ and $\nabla^2\phi = \phi_{;ij}\,dx^i \otimes dx^j$ where

$$
\begin{aligned}
\phi_{;i} &= \partial_{x^i}\phi + \omega_i\phi, \\
\phi_{;ij} &= \partial_{x^j}\partial_{x^i}\phi + \omega_j\partial_{x^i}\phi + \partial_{x^j}(\omega_i\phi) + \omega_j\omega_i\phi - \Gamma_{ij}{}^k\partial_k\phi - \Gamma_{ij}{}^k\omega_k\phi .
\end{aligned}
\tag{17.3.b}
$$

We have $\|\nabla\phi\|^2 = g^{ij}h(\phi_{;i}, \phi_{;j})$. The *rough Laplacian* is given by $D_{g,h,\nabla}\phi := -g^{ij}\phi_{;ij}$. We have $\mathrm{Id} \otimes h : T^*M \otimes V \otimes V \to T^*M$. Consequently, we can regard $h(\nabla\phi, \psi)$ as an element of T^*M. Let $\delta : C^\infty(T^*M) \to C^\infty(M)$ be the coderivative discussed in Lemma 16.4.

The identity $Xh(\phi, \psi) = h(\nabla_X\phi, \psi) + h(\phi, \nabla^*_X\psi)$ defines the dual connection ∇^*. If we compute relative to an orthonormal frame, then the connection 1-form of ∇^* is $-\omega_i^*$. We have that ∇ is unitary if and only if $\nabla^* = \nabla$ or, equivalently, if $\omega_i^* = -\omega_i$.

Lemma 17.16   Let M be a compact manifold. Let ∇ be a connection on (V, h).

1. $(D_{g,h,\nabla}\phi, \psi) = (\nabla\phi, \nabla^*\psi) + \delta(\nabla\phi, \psi)$.
2. $(D_{g,h,\nabla}\phi, \psi)_{L^2} = (\nabla\phi, \nabla^*\psi)_{L^2}$.
3. There exists a constant C_k so that $\|\cdot\|_{k,L^2}^2 \le \varepsilon^2\|\cdot\|_{k+1,L^2} + C_k\varepsilon^{-2}\|\cdot\|_{0,L^2}$ for any $\varepsilon > 0$.

Proof. Let $\{e_1, \ldots, e_m\}$ be a local orthonormal frame for the tangent bundle of M. As the frame for V is orthonormal, the derivatives of the fiber metric h do not enter. We regard

$$(\nabla\phi, \psi) = h(\nabla\phi, \psi)$$

as a 1-form and use Lemma 16.4 to prove Assertion 1 by computing

$$\delta(\nabla\phi, \psi) = -(\phi_i, \psi)_{;i} = -(\phi_{;ii}, \psi) - (\nabla\phi, \nabla^*\psi).$$

Assertion 2 follows by integrating Assertion 1. Assume that ∇ is unitary. We suppose $k = 1$ and use Assertion 2 to see:

$$\|\phi\|_{1,L^2}^2 = \|\nabla\phi\|_{L^2}^2 + \|\phi\|_{L^2}^2 = (D\phi, \phi)_{L^2} + \|\phi\|_{L^2}^2 .$$

Let $\nu > 0$ be given. By Lemma 16.26, $|(D\phi, \phi)_{L^2}| \leq \nu^2 \|D\phi\|_{L^2}^2 + \nu^{-2} \|\phi\|_{L^2}$. We use Lemma 17.14 to estimate $\|D\phi\|_{L^2} \leq C \|\phi\|_{2,L^2}$ and obtain

$$\|\phi\|_{1,L^2}^2 \leq C \|\phi\|_{2,L^2}^2 \nu^2 + (\nu^{-2} + 1) \|\phi\|_{L^2}^2 \,.$$

Choosing $\nu = \nu(\varepsilon)$ suitably then establishes Assertion 3 for $k = 1$; replacing ϕ by $\nabla^j \phi$ and then using induction establishes the result in general. □

17.3.5 OPERATORS OF LAPLACE TYPE.

Let $\phi \in C^\infty(V)$. We choose a local orthonormal frame for V over a coordinate chart $\mathcal{O}$ and regard ϕ as a map from $\mathcal{O}$ to $\mathbb{C}^\ell$. We say that a second-order partial differential operator D on $C^\infty(V)$ is an *operator of Laplace type* if the leading symbol of D is given by the metric tensor, i.e.,

$$D = - \left\{ g^{ij} \operatorname{Id} \partial_{x^i} \partial_{x^j} + A^i \partial_{x^i} + B \right\} \,. \tag{17.3.c}$$

We say that D is *self-adjoint* if $(D\phi, \psi)_{L^2} = (\phi, D\psi)_{L^2}$ for all $\phi, \psi \in C^\infty(V)$. The endomorphisms A^i and B of Equation (17.3.c) are not tensorial. Let ∇ be an arbitrary connection on V and let E be an auxiliary endomorphism of V. We define

$$D_{g,\nabla,E}\phi = -(g^{ij}\phi_{;ij} + E\phi) \,. \tag{17.3.d}$$

Clearly, $D_{g,\nabla,E}$ is an operator of Laplace type; $-g^{ij}\phi_{;ij}$ is the rough Laplacian of Definition 17.15. In fact, every operator of Laplace type can be written in this form; this is the *Bochner formalism* and is closely related to work of Weitzenböck (see Lemma 18.8) and Lichnerowicz (see Lemma 18.21). The metric gives the leading symbol of D, the connection and the Christoffel symbols give the first-order symbol of D, and the metric, connection, and endomorphism give the zero-order symbol of D.

Lemma 17.17   Let D be an operator of Laplace type.

1. There exists a unique connection $\nabla = {}^D\nabla$ on V and a unique endomorphism $E = {}^D E$ on V so that $D = D_{g,\nabla,E}$. Let ω be the connection 1-form of ∇. We have
   (a) $\omega_i = \frac{1}{2} g_{ij}(A^j + g^{k\ell} \Gamma_{k\ell}{}^j \operatorname{Id}_V)$.
   (b) $E = B - g^{ij}(\partial_{x^i}\omega_j + \omega_i \omega_j - \Gamma_{ij}{}^k \omega_k)$.
2. D is self-adjoint if and only if E is self-adjoint and ∇ is unitary.

Proof. We use Equations (17.3.b), (17.3.c), and (17.3.d) to see that:

$$D_{g,\nabla,E} = -g^{ij}(\partial_{x^i}\partial_{x^j} + 2\omega_i \partial_{x^j} - \Gamma_{ij}{}^k \partial_{x^k} + \partial_{x^i}\omega_j + \omega_i \omega_j - \Gamma_{ij}{}^k \omega_k) - E \,.$$

Equating this to $D = -(g^{ij}\partial_{x^i}\partial_{x^j} + A^i \partial_{x^i} + B)$ yields the following relations which we solve for ω and E to prove Assertion 1:

$$A^i = 2g^{ij}\omega_j - g^{k\ell}\Gamma_{k\ell}{}^i \quad \text{and} \quad B = g^{ij}(\partial_{x^i}\omega_j + \omega_i \omega_j - \Gamma_{ij}{}^k \omega_k) + E \,.$$

We apply Lemma 17.16 to both ∇ and ∇^* to compute

$$(D\phi, \psi) = -(D_{g,h,\nabla}\phi, \psi)_{L^2} - (E\phi, \psi)_{L^2} = -(\nabla\phi, \nabla^*\psi)_{L^2} - (E\phi, \psi)_{L^2}$$
$$= -\overline{(\nabla^*\psi, \nabla\phi)_{L^2}} - (\phi, E^*\psi)_{L^2} = -\overline{(D_{g,h,\nabla^*}\psi, \phi)_{L^2}} - (\phi, E^*\psi)_{L^2}$$
$$= -(\phi, (D_{g,h,\nabla^*} + E^*)\psi)_{L^2}.$$

This shows that $D^* = -D_{g,h,\nabla^*} - E^*$. The decomposition of an operator in terms of a rough Laplacian and an endomorphism is unique by Assertion 1. Consequently, D is self-adjoint if and only if $\nabla(D^*) = \nabla(D)^*$ and $E(D^*) = E(D)^*$. Assertion 2 now follows. □

17.4 SPECTRUM OF AN OPERATOR OF LAPLACE TYPE

We conclude Chapter 17 by discussing the spectral theory of self-adjoint operators of Laplace type and of Dirac type on a compact Riemannian manifold. In Sections 17.4.1–17.4.3, we establish the Sobolev Lemma, the Rellich Lemma, and Gårding's inequality for compact manifolds. In Section 17.4.4, we give the spectral resolution of a self-adjoint operator of Laplace type and discuss elliptic regularity. In Section 17.4.5, we define the index of an elliptic complex of Dirac type, and show the index is unchanged by perturbations.

Throughout this section, let (V, h, ∇) be a geometrical vector bundle over a compact Riemannian manifold (M, g). Let $H_k(V)$ and $C^k(V)$ be the completion of $C^\infty(V)$ with respect to the norms $\|\cdot\|_{k,L^2}$ and $\|\cdot\|_{C^k}$, respectively. We omit the dependence on the parameters (g, h, ∇) from the notation in the interests of simplifying the notation since the underlying Banach space structure does not depend on (g, h, ∇) by Lemmas 17.12 and 17.14.

Let ψ_i be a finite partition of unity on M which is subordinate to a cover of M by geodesic balls B_i over which V is trivial. We choose the structures (g_i, h_i, ∇_i) to be flat on B_i. Let $\phi \in C^\infty(V)$. Then we can regard $\psi_i\phi$ as a vector-valued function on $\mathbb{R}^m$ with support in $B_i \subset \mathbb{R}^m$. We defined the norms $\|\cdot\|_{k,L^2(\mathbb{R}^m)}$ on the Schwarz space $\mathcal{S}$ using the Fourier transform. By Lemma 17.8, these are equivalent to the norms $\sum_{j\le k} \|\nabla^j\phi\|_{L^2(V)}$ we have been considering here. We have estimates

$$\varepsilon\|\psi_i\phi\|_{k,C^k(M)} \le \|\psi_i\phi\|_{k,C^k(\mathbb{R}^m)} \le \varepsilon^{-1}\|\psi_i\phi\|_{k,C^k(M)},$$
$$\varepsilon\|\psi_i\phi\|_{k,L^2(V)} \le \|\psi_i\phi\|_{k,L^2(\mathbb{R}^m)} \le \varepsilon^{-1}\|\psi_i\phi\|_{k,L^2(V)}.$$

We sum over i to see that the norms $\|\cdot\|_{k,L^2(\mathbb{R}^m)}$ and $\|\cdot\|_{k,L^2(V)}$ are equivalent as are the norms $\|\cdot\|_{k,C^k(M)}$ and $\|\cdot\|_{k,C^k(\mathbb{R}^m)}$ when dealing with $\phi \in C^\infty(V)$. We therefore blur the distinction between these norms and simply write $\|\cdot\|_{k,C^k}$ and $\|\cdot\|_{k,L^2}$.

17.4.1 THE RELLICH LEMMA FOR M COMPACT [62].

Lemma 17.18 Let (V, h, ∇) be a geometric vector bundle over a compact Riemannian manifold (M, g). The inclusion of $C^\infty(V)$ in $H_k(V)$ extends to define a continuous injective compact linear map $\iota_{H_k, H_j} : H_k(V) \to H_j(V)$ if $k > j \ge 0$.

Proof. We apply the Rellich Lemma for $\mathbb{R}^m$ (see Lemma 17.11). Clearly, $\|_{j,L^2} \leq \|_{k,L^2}$. Consequently, the inclusion of $C^\infty(V)$ in $H_k(V)$ induces a continuous linear map ι_{H_k,H_j} from $H_k(V)$ to $H_j(V)$. Suppose $\iota_{H_k,H_j}\phi = 0$. Let $\phi_n \in C^\infty(V)$ converge to ϕ in $H_k(V)$. Then ϕ_n converge to 0 in $H_j(V)$ and hence in $H_0(V) = L^2(V)$ as well. By Lemma 17.14, pointwise multiplication $m_{\psi_i} : \phi \to \psi_i\phi$ is a continuous linear map from $H_k(V)$ to $H_k(V)$. Consequently, $\psi_i\phi_n$ converges to $\psi_i\phi$ in $H_k(V)$ and $\psi_i\phi_n$ converges to 0 in $L^2(V)$. Let $|\alpha| \leq k$. By Lemma 17.14, ∂_x^α is a continuous linear map from $H^k(V)$ to $L^2(V)$. Set $\phi_{i,\alpha} := \partial_x^\alpha \psi_i\phi$; $\partial_x^\alpha \psi_i\phi_n$ converges to $\phi_{i,\alpha}$ in $L^2(V)$. Let $\xi \in C^\infty(V)$. Since $\psi_i\phi_n$ converges to 0 in $L^2(V)$,

$$(\phi_{i,\alpha}, \xi)_{L^2} = \lim_{n\to\infty} (\partial_x^\alpha \psi_i\phi_n, \xi)_{L^2} = (-1)^{|\alpha|} \lim_{n\to\infty} (\psi_i\phi_n, \partial_x^\alpha \xi)_{L^2} = 0 \,.$$

Since this holds for all $\xi \in C^\infty(V)$, $\phi_{i,\alpha} = 0$ and hence $\|\partial_x^\alpha \psi_i\phi_n\|_{L^2}$ converges to zero in $L^2(V)$ for all $|\alpha| \leq k$. This implies $\|\psi_i\phi_n\|$ converges to zero in $H_k(V)$ so $\psi_i\phi = 0$ in $H_k(V)$. We sum over i to conclude ϕ is zero in $H_k(V)$. Consequently, ι_{H_k,H_j} is injective.

We suppose $\|\phi_n\|_{k,L^2}$ uniformly bounded. This then provides a uniform norm estimate for $\|\psi_i\phi_n\|_{k,L^2}$. We use Lemma 17.11 to extract a subsequence so $\psi_i\phi_n$ converges in $\|_{j,L^2}$ and summing over i shows ϕ_n converges in $\|_{j,L^2}$. □

17.4.2 THE SOBOLEV LEMMA FOR M COMPACT [68].

Lemma 17.19 Let (V, h, ∇) be a geometric vector bundle over a compact Riemannian manifold (M, g). If $k > \frac{1}{2}m + j$, then $\|_{C^j} \leq C_{j,k}\|_{k,L^2}$ and the inclusion of $C^\infty(V)$ in $H_k(V)$ extends to a continuous injective linear map $\iota_{H_k,C^j} : H_k(V) \to C^j(V)$.

Proof. We apply the Sobolev Lemma for $\mathbb{R}^m$ (see Lemma 17.10). By Lemma 17.10,

$$\|\psi_i\phi\|_{g_i,h_i,\nabla_i,C^j(\mathbb{R}^m)} \leq C_i\|\psi_i\phi\|_{g_i,h_i,\nabla_i,k,L^2(\mathbb{R}^m)} \,.$$

By Lemmas 17.12 and 17.14, $\|\psi_i\phi\|_{C^j} \leq C\|\psi_i\phi\|_{k,L^2}$. We use the triangle inequality and Lemma 17.14 to see

$$\|\phi\|_{C^j} = \|\sum_i \psi_i\phi\|_{C^j} \leq \sum_i \|\psi_i\phi\|_{C^j} \quad \text{and} \quad \|\psi_i\phi\|_{k,L^2} \leq C\|\phi\|_{k,L^2} \,.$$

This shows that $\|\phi\|_{C^j} \leq C\|\psi_i\phi\|_{k,L^2}$. Consequently, the identity map of $C^\infty(V)$ induces a well defined continuous linear map ι_{H_k,H_j} from $H_k(V)$ to $C^j(V)$.

Suppose $\iota_{H_k,C^j}(\phi) = 0$. Let $\{\phi_n\} \subset C^\infty(V)$ be a sequence which converges to ϕ in $H_k(V)$. Then the sequence $\{\phi_n\}$ converges to 0 in $C^j(V)$. This implies that the sequence $\{\phi_n\}$ converges to 0 in $H_j(V)$ and, consequently, $\iota_{H_k,H_j}(\phi) = 0$. By Lemma 17.18, this implies $\phi = 0$ in $H_k(V)$ and hence ι_{H_k,C^j} is injective. □

17.4.3 GÅRDING'S INEQUALITY [25].

Lemma 17.20   If D is a self-adjoint operator of Laplace type on a geometric vector bundle (V, h, ∇) over a compact Riemannian manifold (M, g) and if k is a positive integer, then $\|\phi\|_{2k,L^2} \leq C_k (\|\phi\|_{L^2} + \|D^k \phi\|_{L^2})$.

Proof. We generalize Equation (17.2.a) to the setting at hand. We suppose $k = 1$; the general case follows similarly. We use Lemma 17.17 to express $D\phi = -\phi_{;ii} - E\phi$ where we sum over repeated indices relative to a local orthonormal frame for the tangent bundle of M. Lemma 17.16 shows that

$$\|\nabla\phi\|_{L^2}^2 = (\phi_{;i}, \phi_{;i})_{L^2} = |(\phi_{;ii}, \phi)_{L^2}|$$
$$\leq |(D\phi, \phi)_{L^2}| + |(E\phi, \phi)_{L^2}| \leq \|D\phi\|_{L^2}^2 + C_0 \|\phi\|_{L^2}^2 . \tag{17.4.a}$$

In view of Equation (17.4.a), to complete the proof, it suffices to estimate $\|\nabla^2\phi\|_{L^2}$ appropriately. The commutator of covariant differentiation is curvature; the action of curvature is a 0th-order endomorphism which is bounded in L^2. We regard $(\phi_{;ij}, \phi_{;i})e^j$ and $(\phi_{;jj}, \phi_{;i})e^i$ as 1-forms. We integrate by parts using Lemma 17.16 to see

$$\|\nabla^2\phi\|_{L^2} = |(\phi_{;ij}, \phi_{;ij})_{L^2}| = |(\phi_{;ijj}, \phi_i)_{L^2}| \leq |(\phi_{jij}, \phi_i)_{L^2} + C_1 \|\phi\|_{1,L^2}^2$$
$$\leq |(\phi_{;jji}, \phi_{;i})_{L^2}| + C_2 \|\phi\|_{1,L^2}^2 \leq |(\phi_{;jj}, \phi_{;ii})_{L^2}| + C_3 \|\phi\|_{1,L^2}^2$$
$$= |((D + E)\phi, (D + E)\phi)_{L^2}| + C_4 \|\phi\|_{1,L^2}^2$$
$$\leq 2\|D\phi\|_{L^2}^2 + C_5 \|\phi\|_{1,L^2}^2 \leq 3\|D\phi\|_{L^2}^2 + C_6 \|\phi\|_{L^2}^2 . \qquad \square$$

17.4.4 THE SPECTRAL RESOLUTION. We begin with a technical result.

Lemma 17.21   Let D be a self-adjoint operator of Laplace type on a geometric vector bundle (V, h, ∇) over a compact Riemannian manifold (M, g). Let $D_\kappa := D + \kappa \operatorname{Id}$ for $\kappa \in \mathbb{R}$ and let $k \geq 1$. There exists κ_0 so that if $\kappa \geq \kappa_0$, then:

1. $(D_\kappa^k \phi, \phi)_{L^2} \geq \|\phi\|_{L^2}^2$ for all $\phi \in H_{2k}(V)$.
2. $D_\kappa^k : H_{2k}(V) \to L^2(V)$ is a continuous injective linear map.
3. range$\{D_\kappa^k\}$ is closed.
4. $D_\kappa^k : H_{2k}(V) \to L^2(V)$ is surjective.

Proof. Since $H_{2k}(V)$ is the closure of $C^\infty(V)$ with respect to the $\|\cdot\|_{k,L^2}$ norm, we may use Lemma 17.14 to see that D_κ^k is a continuous linear map from $H_{2k}(V)$ to $L^2(V)$. By Lemma 17.18, $H_{2k}(V)$ is a linear subspace of $L^2(V)$. Since $C^\infty(V)$ is dense in $H_k(V)$, we may suppose without loss of generality that ϕ is smooth in the proof of Assertion 1. Suppose first that $k = 1$. Since E is a 0th-order operator, Lemma 17.14 yields $|(E\phi, \phi)|_{L^2} \leq C \|\phi\|_{L^2}^2$ for some constant C. Let $\kappa \geq C + 1$. We use Lemma 17.16 to establish Assertion 1 by estimating.

$$((D + \kappa)\phi, \phi)_{L^2} = (-\phi_{;ii}, \phi)_{L^2} - (E\phi, \phi)_{L^2} + \kappa\|\phi\|_{L^2}^2$$

$$\geq (\phi_{;i}, \phi_{;i})_{L^2} - C(\phi, \phi)_{L^2} + \kappa(\phi, \phi)_{L^2}$$

$$\geq (\kappa - C)(\phi, \phi)_{L^2} \geq \|\phi\|_{L^2}^2.$$

We use Assertion 1 to estimate

$$(D_\kappa^{2\ell}\phi, \phi)_{L^2} = (D_\kappa^\ell\phi, D_\kappa^\ell\phi)_{L^2} \geq 0 \quad \text{for} \quad \ell \geq 1,$$
$$(D_\kappa^{2\ell+1}\phi, \phi)_{L^2} = (D_\kappa D_\kappa^\ell\phi, D_\kappa^\ell\phi)_{L^2} \geq (D_\kappa^\ell\phi, D_\kappa^\ell\phi)_{L^2} \geq 0 \quad \text{for} \quad \ell \geq 0.$$

Let $\mu > 0$. We use the Binomial Theorem to expand

$$(D_{\kappa+\mu}^{2k}\phi, \phi)_{L^2} = \sum_{i+j=2k}\binom{2k}{i}\mu^i(D_\kappa^j\phi, \phi)_{L^2} \geq \mu^{2k}(\phi, \phi)_{L^2}.$$

Consequently, if $\kappa \geq C + 2$, then D_κ^k is a continuous injective linear map from $H_{2k}(V)$ to $L^2(V)$ which proves Assertion 2.

Let $\psi \in \overline{\text{range}\{D_\kappa^k\}}$. Choose $\phi_n \in H_{2k}(V)$ so $D_\kappa^k\phi_n \to \psi$ in L^2. We then have $\{D_\kappa^k\phi\}$ is a Cauchy sequence in L^2. By Assertion 1 and by the interpolation estimate of Lemma 16.26,

$$\|\phi_i - \phi_j\|_{L^2}^2 \leq (D_\kappa^k(\phi_i - \phi_j), \phi_i - \phi_j)_{L^2} \leq \varepsilon^{-2}\|D^k\phi_i - D^k\phi_j\|_{L^2}^2 + \varepsilon^2\|\phi_i - \phi_j\|_{L^2}^2.$$

Taking $\varepsilon < 1$, we conclude $\{\phi_i\}$ is a Cauchy sequence in $L^2(W)$. By Lemma 17.20 (Gårding's inequality) this implies that $\{\phi_i\}$ is a Cauchy sequence in $H_{2k}(W)$ and hence $\phi_i \to \phi_\infty$ in $H_{2k}(W)$. We then have $D_\kappa^k\phi_i \to D_\kappa^k\phi_\infty$ in L^2 and, consequently, $D\phi = \psi$. This shows that $D_\kappa^k(H_{2k}(W))$ is a closed subspace of $L^2(W)$ and establishes Assertion 3.

Since $D_\kappa^k : H_{2k}(W) \to \text{range}\{D_\kappa^k\}$ is 1-1 and onto, and since $\text{range}\{D_\kappa^k\}$ is a closed linear subspace of $L^2(V)$, the Banach–Schauder Theorem (see Theorem 16.20) shows that

$$\mathcal{D}_{\kappa,k} := (D_\kappa^k)^{-1} : \text{range}\{D_\kappa^k\} \to H_{2k}(V)$$

is continuous. The natural map ι from $H_{2k}(V)$ to $L^2(V)$ is a continuous injective compact linear map by Lemma 17.18. Thus, $\iota\mathcal{D}_{\kappa,k}$ is a compact continuous injective linear map. We show that $\iota\mathcal{D}_{\kappa,k}$ is self-adjoint as follows. Let $\phi, \psi \in \text{range}\{D_\kappa^k\}$. Choose $\phi_1, \psi_1 \in H_{2k}(V)$ so that $D_\kappa^k\phi_1 = \phi$ and $D_\kappa^k\psi_1 = \psi$. We have:

$$(\iota\mathcal{D}_{\kappa,k}\phi, \psi)_{L^2} = (\iota(D_\kappa^k)^{-1}D_\kappa^k\psi_1, D_\kappa^k\phi_1)_{L^2} = (\psi_1, D_\kappa^k\phi_1)_{L^2}$$

$$= (D_\kappa^k\psi_1, \phi_1)_{L^2} = (D_\kappa^k\psi_1, \iota\phi_1)_{L^2} = (\psi, \iota\mathcal{D}_{\kappa,k}\phi)_{L^2}.$$

We apply Lemma 16.41 to construct a complete spectral resolution $\{\psi_{n,k}, \mu_{n,k}\}$ for $\mathcal{D}_{\kappa,k}$ on $\text{range}\{D_\kappa^k\}$. Let S be the linear span of the $\psi_{n,k}$; S is a dense subset of $\text{range}\{D_\kappa^k\}$. Consequently, $\mathcal{D}_{\kappa,k}S$ is dense in $H_{2k}(V)$. Since $\iota H_{2k}(V)$ is dense in $L^2(V)$, $\iota\mathcal{D}_{\kappa,k}S$ is dense in $L^2(V)$.

Since $\psi_{n,k}$ is an eigenvector of $\iota\mathcal{D}_{\kappa,k}$ corresponding to a non-zero eigenvalue, we conclude that $\psi_{n,k} = \mu_{n,k}^{-1}\psi_{n,k} \in \iota\mathcal{D}_{\kappa,k}S$ and, consequently, $\iota\mathcal{D}_{\kappa,k}S = S$. This shows that range$\{\mathcal{D}_{\kappa}^k\}$ is dense in $L^2(V)$. Since range$\{\mathcal{D}_{\kappa}^k\}$ is a closed subset of $L^2(V)$, we have that range$\{\mathcal{D}_{\kappa}^k\} = L^2(V)$. This proves Assertion 4. □

Theorem 5.11 of Book II is a consequence of the following result. It generalizes the Fourier series of Example 16.33 to this setting.

Theorem 17.22 Let D be a self-adjoint operator of Laplace type on a geometric vector bundle (V, h, ∇) over a compact Riemannian manifold (M, g).

1. There exists a complete orthonormal basis $\{\phi_n\}$ for $L^2(V)$ so that

$$\phi_n \in C^\infty(V) \quad \text{and} \quad D\phi_n = \lambda_n\phi_n\,.$$

The eigenvalues λ_n tend to $+\infty$; we can choose the indexing so the eigenvalues λ_n are non-decreasing.

2. Choose $0 < k \in \mathbb{Z}$ so $2k > \frac{m}{2}$. Let $0 < \varepsilon < (2k)^{-1}$. There exists $n_0 = n_0(k, m)$ so $\lambda_n \geq n^\varepsilon$ if $n \geq n_0$.

3. For any $0 < k \in \mathbb{Z}$ there exists $k < j(k) \in \mathbb{Z}$ such that $\|\phi_n\|_{C^k} \leq C_k(1 + |\lambda_n|)^{j(k)}$.

4. If $\phi \in L^2(V)$, let $\sigma_n(\phi) := (\phi, \phi_n)_{L^2}$ be the Fourier coefficients. Then $\phi \in C^\infty(V)$ if and only if $\lim_{n\to\infty} \lambda_n^\ell \sigma_n(\phi) = 0$ for all ℓ.

5. If $\phi \in C^\infty(V)$, then the Fourier series $\sum\limits_{n=1}^{\infty} \sigma_n(\phi)\phi_n$ converges to ϕ in $C^\infty(V)$.

6. Let $\phi \in C^\infty(V)$. We may decompose $\phi = \phi_0 + \phi_1$ where ϕ_0 and ϕ_1 are smooth, where $\phi_0 \in \ker\{D\}$, where $\phi_1 \perp \ker\{D\}$, and where $\phi_1 \in D(C^\infty(V))$.

7. If $\phi \in H_2(V)$ and if $D\phi \in C^\infty(V)$, then $\phi \in C^\infty(V)$.

Proof. We can regard ι as the inclusion of $H_{2k}(V)$ in $L^2(V)$; it no longer is needed for book-keeping purposes and we suppress it in the interests of notational simplification henceforth. Choose $\kappa > 0$ so the conclusions of Lemma 17.21 hold. In the proof of Lemma 17.21, we constructed a complete orthonormal basis $\{\psi_{n,k}\}$ for $L^2(V)$ so that $(D_\kappa)^{-1}\psi_{n,k} = \mu_{n,k}\psi_{n,k}$. It is immediate from the definition that $(D_\kappa^k)^{-1} = (D_\kappa)^{-k}$. Consequently, we may set $\psi_n := \psi_{n,1}$, $\mu_n := \mu_{n,1}$, and conclude that $(D_\kappa)^{-k}\psi_n = \mu_n^k\psi_n$. Since $\mu_n^k \neq 0$, $\psi_n \in H_{2k}(V)$ for any k. Lemma 17.19 shows that ψ_n is smooth. It is immediate that $D\psi_n = (D_\kappa - \kappa)\psi_n = \mu_n^{-1} - \kappa$. We set $\lambda_n := \mu_n^{-1} - \kappa$ to prove Assertion 1.

We use an argument due to W. Allard [2] to establish Assertion 2. Suppose first that V is the trivial line bundle so the ϕ_n are simply smooth functions on M. Since the eigenvalues are non-decreasing and tend to ∞, there are only a finite number of negative eigenvalues (or none at all). Thus, we can choose n_0 so that if $n \geq n_0$, then $\lambda_n \geq 1$ and $|\lambda_j| \leq \lambda_n$ for $j \leq n$. Assume $n \geq n_0$ henceforth. Let $E_n := \text{span}\{\phi_1, \ldots, \phi_n\}$. Choose k so $2k > \frac{m}{2}$. If $\phi \in E_n$, then we have

$$\phi = \sum_{j=1}^{n} \gamma_j(\phi)\phi_j, \qquad D^k\phi = \sum_{j=1}^{n} \gamma_j(\phi)\lambda_j^k\phi_j,$$

$$\|D^k\phi\|_{L^2}^2 = \sum_{j=1}^{n} |\lambda_j^k\gamma_j(\phi)|^2 \le \lambda_n^{2k}\sum_{n=1}^{j} |\gamma_j(\phi)|^2 = \lambda_n^{2k}\|\phi\|_{L^2}^2 \,.$$

Consequently, by Lemmas 17.18–17.20, we have that

$$\|\phi\|_\infty^2 \le C\|\phi\|_{2k}^2 \le C\{\|D^k\phi\|_{L^2}^2 + \|\phi\|_{L^2}^2\} \le C\{\lambda_n^{2k} + 1\}\|\phi\|_{L^2}^2 \,.$$

Since $\lambda_n \ge 1$, $\lambda_n^{2k} + 1 \le 2\lambda_n^{2k}$. Thus

$$\|\phi\|_\infty^2 \le 2C\lambda_n^{2k}\|\phi\|_{L^2}^2 \,. \tag{17.4.b}$$

Let $\dim(V) = \ell$. Fix a point x of M and choose a local orthonormal frame for W near x to express $\phi_j = (\phi_{j,1}, \dots, \phi_{j,\ell})$ as a vector-valued function. Fix a coordinate $1 \le \mu \le \ell$. Let

$$c_j = \bar{\phi}_{j,\mu}(x) \quad \text{and} \quad \phi = c_1\phi_1 + \dots + c_n\phi_n \,.$$

We use Equation (17.4.b) to estimate

$$\left\{\sum_{j=1}^{n} |\phi_{j,\mu}(x)|^2\right\}^2 = |\{c_1\phi_1(x) + \dots + c_n\phi_n(x)\}_\mu|^2 \le \|\phi(x)\|^2 \le \|\phi\|_\infty^2$$

$$\le 2C\lambda_n^{2k}\|\phi\|_{L^2}^2 = 2C\lambda_n^{2k}\sum_{j=1}^{n} |\phi_{j,\mu}(x)|^2 \,.$$

Consequently, $\displaystyle\sum_{j=1}^{n} |\phi_{j,\mu}(x)|^2 \le 2C\lambda_n^{2k}$ and hence

$$\sum_{j=1}^{n} \|\phi_j(x)\|^2 = \sum_{j=1}^{n}\sum_{\mu=1}^{\ell} |\phi_{j,\mu}(x)|^2 \le 2\ell C\lambda_n^{2k} \,.$$

Integrating this estimate over M yields

$$n = \sum_{j=1}^{n} \|\phi_j\|_{L^2}^2 = \int_M \sum_{j=1}^{n} \|\phi_j(x)\|^2 |\,\mathrm{dvol}\,|(x) \le 2\ell C\lambda_n^{2k}\int_M |\,\mathrm{dvol}\,|(x) = 2\ell C\lambda_n^{2k}\,\mathrm{vol}(M) \,.$$

Consequently, if $n \ge n_0$, $\lambda_n \ge n^{\frac{1}{2k}}\{2\ell C\}^{\frac{1}{2k}}$. Assertion 3 now follows since $0 < \varepsilon < \frac{1}{2k}$ implies $n^{\frac{1}{2k}}\{2\ell C\}^{\frac{1}{2k}} \ge n^\varepsilon$ for large n.

We apply Gårding's inequality (see Lemma 17.20) to see

$$\|\phi_n\|_{2j,L^2}^2 \le C\left\{\|D^j\phi_n\|_{L^2}^2 + \|\phi_n\|_{L^2}^2\right\} = C(|\lambda_n^j|^2 + 1) \le C(|\lambda_n| + 1)^{2j} \,.$$

Choose j so $j > \frac{1}{2}m + k$. We apply Lemma 17.19 to estimate $\|\cdot\|_{C^k} \leq C\|\cdot\|_{j,L^2}$. Assertion 3 now follows.

We now prove Assertion 4. Let $\phi \in C^\infty(V)$. We estimate that

$$|\lambda_n^{k+1}(\phi, \phi_n)_{L^2}| = |(\phi, D^{k+1}\phi_n)_{L^2}| = |(D^{k+1}\phi_n, \phi)_{L^2}| \leq \|D^{k+1}\phi\|_{L^2} \|\phi_n\| \leq C_k$$

for all n. Thus, $|\lambda_n^k \sigma_n| \leq C_k \lambda_n^{-1} \to 0$ so $\lim_{n\to\infty} \lambda_n^k |\sigma_n(\phi)| = 0$. Conversely, suppose that $\lim_{n\to\infty} \lambda_n^\ell |\sigma_n(\phi)| = 0$ for all ℓ. Choose ℓ so $(j(k) - \ell)\varepsilon < -2$. We may estimate

$$\sum_{n=1}^N \|\sigma_n(\phi)\phi_n\|_{k,L^2} \leq C \sum_{n=1}^N |\sigma_n(\phi)|\,|\lambda_n^{j(k)}| \leq C \sum_{n=1}^N \lambda_n^{j(k)-\ell} \leq \sum_{n=1}^N n^{(j(k)-\ell)\varepsilon} < \infty.$$

This implies the series $\sum_n \sigma_n(\phi)\phi_n$ converges to ϕ in the C^∞ topology and hence ϕ is smooth. Assertions 4 and 5 now follow.

Expand $\phi \in C^\infty(V)$ in a Fourier series $\phi = \sum_n \sigma_n(\phi)\phi_n$. By Assertion 5, the Fourier series converges in the C^∞ topology. Since $\lim_{n\to\infty} \lambda_n = \infty$, there are a finite number of zero eigenvalues and $\phi_0 := \sum_{\lambda_n=0} \sigma_n(\phi)\phi_n \in \ker\{D\}$ is smooth. Furthermore,

$$\phi_1 = \phi - \phi_0 = \sum_{\lambda_n \neq 0} \sigma_n(\phi)\phi_n \in \ker\{D\}^\perp.$$

Finally, since $\lim_{n\to\infty} \lambda_n = \infty$, there exists $\varepsilon > 0$ so $\lambda_n \neq 0$ implies $|\lambda_n| \geq \varepsilon$. We set

$$\psi_1 = \sum_{\lambda_n \neq 0} \lambda_n^{-1} \sigma_n(\phi)\phi_n.$$

We have that $|\sigma_n(\psi)| = |\lambda_n^{-k-1}|$ for any k and hence by Assertion 5, ψ_1 is smooth. Assertion 6 follows since $D\psi_1 = \phi_1$.

Let $\phi \in H_2(V)$. Suppose that $\psi := D\phi$ is smooth. If n is large, then λ_n is positive and we derive Assertion 7 from Assertion 4 by estimating for any k and for $n \geq n_0$

$$\sigma_n(\phi) = (\phi, \phi_n)_{L^2} = \lambda_n^{-1}(\phi, D\phi_n)_{L^2} = \lambda_n^{-1}(D\phi, \phi_n)_{L^2} = \lambda_n^{-1}(\psi, \phi_n)_{L^2} \leq C(\lambda_n^{-k}). \quad \square$$

Assertion 7 of Theorem 17.22 is a global result. There is a corresponding local result of elliptic regularity.

Theorem 17.23   Let D be a self-adjoint operator of Laplace type on a geometric vector bundle (V, h, ∇) over a compact Riemannian manifold (M, g). Let $\mathcal{O}$ be an open subset of M. If ϕ belongs to $H_2(V)$ and if $D\phi|_{\mathcal{O}} \in C^\infty(\mathcal{O})$, then $\phi|_{\mathcal{O}} \in C^\infty(\mathcal{O})$.

Proof. The subspaces $H_k(V)$ embed in $L^2(V)$ as a nested decreasing sequence

$$L^2(V) \supset H_1(V) \supset H_2(V) \supset \cdots.$$

Fix a point P of $\mathcal{O}$. Choose a *plateau function* ϑ which is identically near 1 near P and which has compact support in $\mathcal{O}$. Let $m_\vartheta : \phi \to \vartheta \cdot \phi$; this is a continuous linear map from $H_k(V)$ to $H_k(V)$ by Lemma 17.14. We assumed $\phi \in H_2(V)$ so $m_\vartheta \phi \in H_2(V)$. Suppose inductively we have shown that $m_\vartheta \phi \in H_k(V)$. By assumption, $m_\vartheta D\phi \in C^\infty(V) \subset H_{k+1}(V)$. We may express

$$Dm_\vartheta \phi = m_\vartheta D\phi + [D, m_\vartheta]\phi.$$

The commutator $[D, m_\vartheta]$ is a first-order partial differential operator. Consequently,

$$[D, m_\vartheta]\phi \in H_{k-1}(V) \quad \text{so} \quad Dm_\vartheta \phi \in H_{k-1}(V).$$

The argument given to prove Theorem 17.22 extends to show $m_\vartheta \phi \in H_{k+1}(V)$ so we have a net gain of 1 derivative. Consequently, $m_\vartheta \phi \in H_k(V)$ for all k and Lemma 17.19 implies $m_\vartheta \phi$ belongs to $C^\infty(V)$. Since $\vartheta \equiv 1$ near P, this shows that ϕ is smooth near P.   □

17.4.5 THE HODGE DECOMPOSITION THEOREM [39]. Let (M, g) be a compact Riemannian manifold. For $0 \leq i \leq \ell$, we assume given Hermitian vector bundles (V_i, h_i) and first-order partial differential operators $d_i : C^\infty(V_i) \to C^\infty(V_{i+1})$ which satisfy $d_i^2 = 0$. Assume the associated second-order operators $\Delta_i := \delta_i d_i + d_{i-1}\delta_{i-1}$ are operators of Laplace type on $C^\infty(V_i)$ where $\delta_i : C^\infty(V_{i+1}) \to C^\infty(V_i)$ is the formal adjoint of d_i. The collection $\mathcal{E} := \{V_i, h_i, d_i\}$ is then called an *elliptic complex of Dirac type*. Define the associated cohomology groups by setting

$$H^p(\mathcal{E}) := \frac{\ker\{d_i : C^\infty(V_i) \to C^\infty(V_{i+1})\}}{\text{range}\{d_{i-1} : C^\infty(V_{i-1}) \to C^\infty(V_i)\}}.$$

Theorem 17.24   Let $\mathcal{E}$ be an elliptic complex of Dirac type over a compact Riemannian manifold (M, g). Let $\phi_i \in C^\infty(V_i)$.

1. $\Delta_i \phi_i = 0$ if and only if $d_i \phi_i = 0$ and $\delta_{i-1}\phi_i = 0$.
2. There is an orthogonal direct sum decomposition

$$C^\infty(V_i) = \ker\{\Delta_i\} \oplus d_{i-1}C^\infty(V_{i-1}) \oplus \delta_i C^\infty(V_{i+1}).$$

3. The inclusion of $\ker\{\Delta_i\}$ into $\ker\{d_i\}$ induces an isomorphism from $\ker\{\Delta_i\}$ to $H^p(\mathcal{E})$. Consequently, $\dim(H^p(\mathcal{E})) = \dim(\ker\{\Delta_i\}) < \infty$.

Proof. Since $\Delta_i = \delta_i d_i + d_{i-1}\delta_{i-1}$, if $\phi_i \in \ker\{d_i\} \cap \ker\{\delta_{i-1}\}$, then $\Delta_i \phi_i = 0$. Conversely, suppose $\Delta_i \phi_i = 0$. By Lemma 17.14, the natural map from $C^\infty(V_i)$ to $L^2(V_i)$ is injective. We show $d_i \phi_i = 0$ and $\delta_{i-1}\phi_i = 0$ and establish Assertion 1 by computing:

$$
\begin{aligned}
0 &= (\Delta_i \phi_i, \phi_i)_{L^2} = (\delta_i d_i \phi_i, \phi_i)_{L^2} + (d_{i-1}\delta_{i-1}\phi_i, \phi_i)_{L^2} \\
&= (d_i \phi_i, d_i \phi_i)_{L^2} + (\delta_{i-1}\phi_i \delta_{i-1}\phi_i)_{L^2} = \|d_i \phi_i\|_{L^2}^2 + \|\delta_{i-1}\phi_i\|_{L^2}^2.
\end{aligned}
$$

By Theorem 17.22,

$$C^\infty(V_i) = \ker\{\Delta_i\} + \Delta_i C^\infty(V_i) = \ker\{\Delta_i\} + d_{i-1}C^\infty(V_{i-1}) + \delta_k C^\infty(V_{i+1}). \quad (17.4.c)$$

Let $\psi_i \in \ker\{\Delta_i\} = \ker\{d_i\} \cap \ker\{\delta_{i-1}\}$, let $\psi_{i+1} \in C^\infty(V_{i+1})$, and let $\psi_{i-1} \in C^\infty(V_{i-1})$. We show that Equation (17.4.c) is an orthogonal decomposition and thereby establish Assertion 2 by computing:

$$(\psi_i, \delta_i \psi_{i+1})_{L^2} = (d_i \psi_i, \psi_{i+1})_{L^2} = 0,$$
$$(\psi_i, d_{i-1} \psi_{i-1})_{L^2} = (\delta_{i-1} \psi_i, \psi_{i-1})_{L^2} = 0,$$
$$(d_{i-1} \psi_{i-1}, \delta_i \psi_{i+1})_{L^2} = (d_i d_{i-1} \psi_{i-1}, \psi_{i+1})_{L^2} = 0.$$

If $d\phi = 0$, let $[\phi] \in H^i(\mathcal{E})$ denote the corresponding cohomology class. Let ι be the map from $\ker\{\Delta_i\}$ to $H^i(\mathcal{E})$ which sends ϕ to $[\phi]$. Let $\phi \in \ker\{\Delta_i\}$. Suppose $\iota(\phi) = 0$. This implies $\phi = d\psi_{i-1}$ for some $\psi_{i-1} \in C^\infty(V_{i-1})$. Since $\ker\{\Delta_i\} \perp dC^\infty(V_{i-1})$, this implies $\phi = 0$. Thus, ι is injective. Let $\phi \in \ker\{d\}$. Decompose $\phi = \phi_i + d_{i-1}\phi_{i-1} + \delta_i\phi_{i+1}$. Since $d\phi = 0$, $d\delta_i\phi_{i+1} = 0$. We show $\delta_i\phi_{i+1} = 0$ by computing

$$0 = (d_i \delta_i \phi_{i+1}, \phi_{i+1})_{L^2} = (\delta_i \phi_{i+1}, \delta_i \phi_{i+1})_{L^2} = \|\delta_i \phi_{i+1}\|^2.$$

This shows $\phi = \phi_i + d\psi_{i-1}$ so $[\phi] = [\phi_0]$ and ι is surjective.   □

By Theorem 17.22, $\ker\{\Delta_i\}$ is finite-dimensional. We define

$$\mathrm{Index}(\mathcal{E}) := \sum_{i=0}^{k} \dim(\ker\{\Delta_i\}).$$

We can *wrap up the elliptic complex* to define a two-term complex of Dirac type. Define

$$W_+ := \oplus_p V_{2p}, \qquad \Delta_+ = \oplus_p \Delta_{2p}, \qquad A_+ := \oplus_p(d_{2p} + \delta_{2p-1}),$$
$$W_- := \oplus_p V_{2p+1}, \qquad \Delta_- = \oplus_p \Delta_{2p+1}, \qquad A_- := \oplus_p(d_{2p+1} + \delta_{2p}).$$

The operators $\Delta_\pm$ are self-adjoint operators of Laplace type. Since

$$A_- = A_+^*, \qquad \Delta_+ = A_+^* A_+, \qquad \Delta_- = A_+ A_+^*,$$

we have that $\mathcal{F} := (W_+, W_-, A_+)$ is an elliptic complex of Dirac type. It is immediate from the definition that

$$\mathrm{Index}(\mathcal{F}) = \dim(\ker\{\Delta_+\}) - \dim(\ker\{\Delta_-\})$$
$$= \sum_p \{\dim(\ker\{\Delta_{2p}\}) - \dim(\ker\{\Delta_{2p+1}\})\} = \mathrm{Index}(\mathcal{E}).$$

Lemma 17.25   Let $A_+ : C^\infty(W_+) \to C^\infty(W_-)$ define an elliptic complex $\mathcal{E}$ of Dirac type over a compact Riemannian manifold (M, g). Then $A_+ : H_1(W_+) \to L^2(W_-)$ is a Fredholm operator and $\mathrm{Index}(A_+) = \mathrm{Index}(\mathcal{E})$.

Proof. We apply Theorems 17.22 and 17.24. We have $\ker\{\Delta_+\} = \ker\{A_+\}$ is finite-dimensional. Let π_+ be orthogonal projection in L^2 on $\ker\{A_+\}$; π_+ is a continuous linear map from $H_1(W_+)$ to $\ker\{A_+\}$; set

$$\tilde{H}_1(W_+) := \ker\{\pi_+\} = H_1(W_+) \cap \ker\{A_+\}^\perp \,.$$

We observe that $C^\infty(W_+) \cap \ker\{A_+\}^\perp$ is dense in $\tilde{H}_1(W_+)$. Let $\varepsilon > 0$ be the smallest non-zero eigenvalue of Δ_+. If $\phi \in C^\infty(W_+) \cap \ker\{A_+\}^\perp$, then

$$(A_+\phi, A_+\phi)_{L^2(W_-)} = (\Delta_+\phi, \phi)_{L^2(W_+)} \geq \varepsilon(\phi, \phi)_{L^2(W_+)} \,. \tag{17.4.d}$$

We compute relative to a local orthonormal frame for TM to express $A_+\phi = \gamma^i\phi_{;i} + \gamma$ for suitably chosen elements of $\mathrm{Hom}(W_+, W_-)$. The condition that A_+ is an operator of Laplace type means that $(\gamma^i)^*\gamma^j = \delta^{ij}$. Let $\phi \in C^\infty(W_+) \cap \ker\{A_+\}^\perp$. We use Equation (17.4.d) to see

$$\begin{aligned}
\|\phi\|_{1,L^2}^2 &= \|\nabla\phi\|_{L^2(W_+)}^2 + \|\phi\|_{L^2(W_+)}^2 = ((c^i)^*c^j\phi_j, \phi_i)_{L^2(W_+)} + \|\phi\|_{L^2(W_+)}^2 \\
&= (c^j\phi_{;j}, c^i\phi_{;i})_{L^2(W_-)} + \|\phi\|_{L^2(W_+)}^2 = \|(A_+ - \gamma)\phi\|_{L^2(W_-)}^2 + \|\phi\|_{L^2(W_+)}^2 \\
&\leq \|A_+\phi\|_{L^2(W_-)}^2 + C_1\|\phi\|_{L^2(W_+)}^2 \leq C_2\|A_+\phi\|_{L^2(W_-)}^2 \,.
\end{aligned}$$

This shows that $\mathrm{range}\{A_+ : H_2(W_+) \to L^2(W_-)\}$ is a closed linear subspace of $L^2(W_-)$. Let $\{\psi_{n,-}, \lambda_{n,-}\}$ be a complete spectral resolution of Δ_-. Then $\psi_{n,-} \in \mathrm{range}\{A_+\}$ if $\lambda_{n,-} \neq 0$. Thus if $\phi \in \mathrm{range}\{A_+\}^\perp$, then $(\phi, \phi_{n,-})_{L^2(W_-)} = 0$ for $\lambda_{n,-} \neq 0$ so $\phi \in \ker\{\Delta_-\}$. Since

$$\ker\{\Delta_-\} \perp \mathrm{range}\{A_+\}\,,$$

we have $\mathrm{coker}\{A_+\} = \ker\{\Delta_-\}$. This shows A_+ is Fredholm and, furthermore, that

$$\mathrm{Index}(A_+) = \dim(\ker\{\Delta_+\}) - \dim(\ker\{\Delta_-\}) = \mathrm{Index}(\mathcal{E})\,. \qquad \square$$

Lemmas 16.42 and 17.25 yield the following result.

Lemma 17.26   Let V be a vector bundle over a compact manifold M. Let g_t be a smooth one-parameter family of Riemannian metrics on M, let $h_{i,t}$ be a smooth one-parameter family of fiber metrics on smooth vector bundles V_i over M, and let $d_{i,t}$ be a smooth one-parameter family of first-order partial differential operators mapping $C^\infty(V_i)$ to $C^\infty(V_{i+1})$ so that $\mathcal{E}_t = \{V_i, d_{i,t}\}$ is an elliptic complex of Dirac type. Then $\mathrm{Index}(\mathcal{E}_t)$ is independent of the parameter t.

CHAPTER 18

Potential Theory

We shall discuss work of the following mathematicians, among others, in Chapter 18:

M. Atiyah
(1929-2019)

E. Beltrami
(1835–1900)

R. Bott
(1923-2005)

W. Clifford
(1845-1879)

A. Lichnerowicz
(1915–1998)

L. Nirenberg
(1925–2020)

H. Poincaré
(1854–1912)

G. Riemann
(1826-1866)

G. Roch
(1839–1866)

J. P. Serre
(1926–)

E. Stiefel
(1909-1978)

R. Weitzenböck
(1885–1955)

H. Whitney
(1907–1989)

In Section 18.1, we introduce Clifford algebras and discuss operators of Dirac type. In Section 18.2, we discuss the de Rham complex. In Section 18.3, we treat spinors. In Section 18.4, we begin our treatment of the Dolbeault complex. In Section 18.5, we continue our discussion with a treatment of Serre duality and the Kodaira Vanishing Theorem for the complex Laplacian.

We shall always assume given a positive definite inner product on any vector bundle (resp. vector space). This lets one identify the bundle (resp. vector space) with the dual bundle (resp. dual vector space). We shall for the most part work with orthonormal frames (resp. bases) and use Roman indices. In this setting, we shall blur the distinction between upper and lower indices and simply write all Roman indices down and sum over repeated Roman indices. Occasionally, when using coordinate frames for the tangent and cotangent bundles, we will use Greek indices and preserve the distinction between upper and lower indices.

18.1   OPERATORS OF DIRAC TYPE

In Section 18.1.1, we define the exterior algebra $\Lambda(V)$ and the Clifford algebra $\mathrm{Cliff}(V, h)$ where V is a finite-dimensional vector space equipped with a positive definite inner product h. In Section 18.1.2, we express the Hodge $\star$ operator in terms of Clifford multiplication. In Section 18.1.3, we discuss the representation theory of the complex Clifford algebras. In Section 18.1.4, we extend these notions to the context of vector bundles over a Riemannian manifold (M, g), and in Section 18.1.5, we discuss operators of Dirac type.

18.1.1 THE EXTERIOR ALGEBRA AND THE CLIFFORD ALGEBRA. We refer to Atiyah, Bott and Shapiro [6] for further details concerning this material. We work pointwise. Let h be a positive definite inner product on an m-dimensional real vector space V. Let

$$\mathcal{B} = \{e_1, \ldots, e_m\}$$

be an orthonormal basis for V. The *exterior algebra* $\Lambda(V) = \oplus_p \Lambda^p(V)$ and the *Clifford algebra* $\mathrm{Cliff}(V, h)$ are the universal unital real algebras generated by V subject to the relations

$$v \wedge w + w \wedge v = 0 \quad \text{and} \quad v * w + w * v = -2h(u, v)\mathbb{1},$$

where $\mathbb{1}$ is the unit of these algebras and where v and w are arbitrary elements of V. More precisely, let $\mathcal{T}(V) = \oplus_p \otimes^p V$ be the complete tensor algebra generated by V. Let $\mathcal{I}_\Lambda$ and $\mathcal{I}_{\mathrm{Cliff}}$ be the two-sided ideals of $\mathcal{T}(V)$ generated by all elements of the form $v \otimes w + w \otimes v$ and $v \otimes w + w \otimes v + 2h(v, w)\mathbb{1}$, respectively. Then

$$\Lambda(V) := \mathcal{T}(V)/\mathcal{I}_\Lambda \quad \text{and} \quad \mathrm{Cliff}(V, h) := \mathcal{T}(V)/\mathcal{I}_{\mathrm{Cliff}}. \tag{18.1.a}$$

Let $I = \{1 \leq i_1 < \cdots < i_p \leq m\}$ be an increasing collection of indices. Set

$$|I| := p, \quad e_{\Lambda, I} := e_{i_1} \wedge \cdots \wedge e_{i_p} \quad \text{and} \quad e_{\mathrm{Cliff}, I} := e_{i_1} * \cdots * e_{i_p}.$$

Lemma 18.1 $\{e_{\Lambda,I}\}$ is a basis for $\Lambda(V)$ and $\{e_{\mathrm{Cliff},I}\}$ is a basis for $\mathrm{Cliff}(V,h)$.

Proof. It is immediate from the defining relationships that $\{e_{\Lambda,I}\}$ spans $\Lambda(V)$ and that $\{e_{\mathrm{Cliff},I}\}$ spans $\mathrm{Cliff}(V,h)$. We first show that $\{e_{\Lambda,I}\}$ is a linearly independent subset of $\Lambda(V)$. Let $\mathfrak{D}(v_1,\ldots,v_m) := \det h(e_i,v_j)$ define a multi-linear map from $V \times \cdots \times V$ to $\mathbb{R}$. Extend $\mathfrak{D}$ to a linear map from $\otimes^m V$ to $\mathbb{R}$. If we interchange two rows of a matrix, we change the sign of the determinant. Consequently, $\mathfrak{D}$ vanishes on the defining ideal $\mathcal{I}_\Lambda$ and extends to a well defined map $\mathfrak{D} : \Lambda^m(V) \to \mathbb{R}$. Since $\mathfrak{D}(e_1 \wedge \cdots \wedge e_m) = 1$, $e_1 \wedge \cdots \wedge e_m \neq 0$.

Suppose we have a non-trivial dependence relation $\sum_I a_I e_{\Lambda,I} = 0$. We argue for a contradiction. Choose I so that $a_I \neq 0$. By reordering the elements of the basis $\mathcal{B}$, we may assume $I = \{1,\ldots,p\}$. Let J be the complementary set, $J := \{p+1,\ldots,m\}$. We extend $\mathfrak{D}$ to be zero on Λ^p for $p < m$. Then $\mathfrak{D}\{e_{\Lambda,K} \wedge e_{\Lambda,J}\} = \delta_{IK}$. Assertion 1 follows from the following contradiction:

$$0 = \mathcal{D}\sum_K a_K e_{\Lambda,K} \wedge e_{\Lambda,K} = a_I \neq 0 \,.$$

Since $\{e_\Lambda^I\}$ is a basis for $\Lambda(V)$, we may define an inner product $\Lambda(V)$ by setting

$$h_{\Lambda,\mathcal{B}}(e_{\Lambda,I},e_{\Lambda,J}) = \delta_{IJ} \,. \tag{18.1.b}$$

Let $\mathrm{ext}(v)\Theta := v \wedge \Theta$ be *exterior multiplication*, and let $\mathrm{int}_\mathcal{B}(v)$ be *interior multiplication*; this is the dual map relative to the inner product $h_{\Lambda,\mathcal{B}}$. It is immediate that:

$$\mathrm{ext}(e_1)(e_{i_1} \wedge \cdots \wedge e_{i_p}) = \left\{ \begin{array}{ll} 0 & \text{if } i_1 = 1 \\ e_1 \wedge e_{i_1} \wedge \cdots \wedge e_{i_p} & \text{if } i_1 > 1 \end{array} \right\},$$

$$\mathrm{int}_\mathcal{B}(e_1)(e_{i_1} \wedge \cdots \wedge e_{i_p}) = \left\{ \begin{array}{ll} e_{i_2} \wedge \cdots \wedge e_{i_p} & \text{if } i_1 = 1 \\ 0 & \text{if } i_1 > 1 \end{array} \right\}.$$

Let $c_{\Lambda,\mathcal{B}}(v) := \mathrm{ext}(v) - \mathrm{int}_\mathcal{B}(v)$; this is called *Clifford multiplication*. We obtain the *Clifford identity*:

$$c_{\Lambda,\mathcal{B}}(v)c_{\Lambda,\mathcal{B}}(w) + c_{\Lambda,\mathcal{B}}(w)c_{\Lambda,\mathcal{B}}(v) = -2h_{\Lambda,\mathcal{B}}(w,v)\,\mathrm{Id} \,. \tag{18.1.c}$$

Because $c_{\Lambda,\mathcal{B}}$ vanishes on the ideal $\mathcal{I}_{\mathrm{Cliff}}$, $c_{\Lambda,\mathcal{B}}$ extends to a unital algebra morphism from $\mathrm{Cliff}(V,h)$ to the algebra $\mathrm{End}(\Lambda(V))$ of endomorphisms of the exterior algebra. Let

$$\Psi_\mathcal{B} : \vartheta \to c_{\Lambda,\mathcal{B}}(\vartheta)\mathbb{1} \quad \text{for} \quad \vartheta \in \mathrm{Cliff}(V,h) \,. \tag{18.1.d}$$

Since $\Psi_\mathcal{B} e_{\mathrm{Cliff},I} = e_{\Lambda,I}$, and since $\{e_{\Lambda,I}\}$ is a linearly independent set, $\{e_{\mathrm{Cliff},I}\}$ is a linearly independent set. $\qquad\square$

The ideal $\mathcal{I}_\Lambda$ is homogeneous. Consequently, $\Lambda(V)$ decomposes into homogeneous pieces. The ideal $\mathcal{I}_{\mathrm{Cliff}}$ is not homogeneous but rather is $\mathbb{Z}_2$ graded so

$$\mathrm{Cliff}(V,h) = \mathrm{Cliff}^+(V) \oplus \mathrm{Cliff}^-(V)$$

decomposes into even (+) and odd (−) pieces. We then have $\{e_{\Lambda,I}\}$ for $|I| = p$ is a basis for $\Lambda^p(V)$, $\{e_{\mathrm{Cliff},I}\}$ for $|I|$ even is a basis for $\mathrm{Cliff}^+(V)$, and $\{e_{\mathrm{Cliff},I}\}$ for $|I|$ odd is a basis for $\mathrm{Cliff}^-(V)$.

We define the *graded transpose* on the tensor algebra by setting

$$\{e_{i_1} \otimes \cdots \otimes e_{i_p}\}^t = (-1)^p e_{i_p} \otimes \cdots \otimes e_{i_1}.$$

The map $\omega \to \omega^t$ is a unital anti-algebra morphism of the tensor algebra which preserves the ideal $\mathcal{I}_{\mathrm{Cliff}}$. Consequently, the transpose extends to a unital anti-algebra morphism of $\mathrm{Cliff}(V,h)$. Let $h_{\Lambda,\mathcal{B}}$ be the inner product on the exterior algebra defined by Equation (18.1.b); we use the isomorphism $\Psi_\mathcal{B}$ of Equation (18.1.d) to define a inner product $h_{\mathrm{Cliff},\mathcal{B}}$ on $\mathrm{Cliff}(V,h)$ so that $h_{\mathrm{Cliff},\mathcal{B}}(e_{\mathrm{Cliff},I}, e_{\mathrm{Cliff},J}) = \delta_{I,J}$. If $\phi \in \mathrm{Cliff}(V,h)$, let $m_L(\phi) : \psi \to \phi * \psi$ be left Clifford multiplication; m_L is a unital algebra morphism from $\mathrm{Cliff}(V,h)$ to $\mathrm{End}(\mathrm{Cliff}(V,h))$.

Lemma 18.2

1. The inner products $h_{\Lambda,\mathcal{B}}$ and $h_{\mathrm{Cliff},\mathcal{B}}$ are independent of $\mathcal{B}$.

2. Let $\mathcal{Z}(V,h)$ be the center of $\mathrm{Cliff}(V,h)$.

   (a) If m is even, then $\mathcal{Z}(V,h) = \mathbb{R} \cdot 1$.

   (b) If m is odd, then $\mathcal{Z}(V,h) = \mathbb{R} \cdot 1 \oplus \{e_1 * \cdots * e_m\} \cdot \mathbb{R}$.

Proof. The bilinear form $\tilde{h}(\phi, \psi) := \mathrm{Tr}\{m_L(\phi * \psi^t)\}$ is invariantly defined. If $J \neq K$, then there exists L non-trivial so

$$e_{\mathrm{Cliff},J} * (e_{\mathrm{Cliff},K})^t = \pm e_{\mathrm{Cliff},L}$$

and, consequently, $m_L(e_{\mathrm{Cliff},J} * (e_{\mathrm{Cliff},K})^t)$ permutes without fixed points and modulo sign the basis $\{e_{\mathrm{Cliff},I}\}$ of $\mathrm{Cliff}(V,h)$. This shows $\tilde{h}(e_{\mathrm{Cliff},J}, e_{\mathrm{Cliff},K}) = 0$ if $J \neq K$. On the other hand, if $J = K$, then $e_{\mathrm{Cliff},J} * (e_{\mathrm{Cliff},K})^t = 1$ so $\tilde{h}(e_I, e_I) = \dim(\mathrm{Cliff}(V,h)) = 2^m$. Consequently, $h_\mathcal{B} = 2^{-m}\tilde{h}$ is independent of $\mathcal{B}$. This establishes Assertion 1.

Let $\theta = \sum_I a_I e_{\mathrm{Cliff},I} \in \mathcal{Z}(V,h)$. Then $e_i \theta = \theta e_i$ implies $e_i \theta e_i = -\theta$ for all i. If $|I| > 0$ and $a_I \neq 0$, then $e_i * e_{\mathrm{Cliff},I} * e_i = -e_{\mathrm{Cliff},I}$. Express $I = \{1 \leq i_1 < \cdots < i_p \leq m\}$. If p is even, then $e_{i_1} * e_{\mathrm{Cliff},I} * e_{i_1} = e_{\mathrm{Cliff},I}$ which is false. If p is odd, and some index j does not belong to I, then $e_j * e_{\mathrm{Cliff},I} * e_j = e_{\mathrm{Cliff},I}$ which is false. Thus, the only possibility is that either $\theta = a1$ or that $\theta = a_0 1 + a_1 e_1 * \cdots * e_m$ for m odd. Assertion 2 follows since such an element is in fact central in $\mathrm{Cliff}(V,h)$. $\square$

By Lemma 18.2, $\text{int}_{\mathcal{B}}$ and $c_{\Lambda,\mathcal{B}}$ are independent of $\mathcal{B}$ as well. We therefore drop the dependence on $\mathcal{B}$ from the notation henceforth. Furthermore, since h_Λ and h_{Cliff} extend the metric from V to $\Lambda(V)$ or $\text{Cliff}(V, h)$, respectively, we also drop the dependence on Λ and Cliff and simply speak of h; the map Ψ of Equation (18.1.d) is then an isometry; we caution that Ψ is not an algebra morphism.

18.1.2 THE HODGE $\star$ OPERATOR. Let (V, h) be an oriented real vector space which is equipped with a positive definite inner product h. Choose an oriented orthonormal basis

$$\mathcal{B} = \{e_1, \ldots, e_m\}$$

for V and let $\text{orn} := e_1 \wedge \cdots \wedge e_m \in \Lambda^m(V)$. The *Hodge $\star$ operator* (see Section 5.2 of Book II) is the collection of linear isometries $\star_p$ mapping $\Lambda^p(V)$ to $\Lambda^{m-p}(V)$ characterized by the property that $h(\theta, \eta)\text{orn} = \theta \wedge \star_p\eta$. We can use the linear isomorphism between $\Lambda(V)$ and $\text{Cliff}(V, h)$ defined by Equation (18.1.d) to regard $\text{orn} = e_1 * \cdots * e_m$ as element of the Clifford algebra. One then has (see Lemma 5.9 of Book II) that:

Lemma 18.3 $\star_p = (-1)^{\frac{1}{2}(2m-p+1)p} c_\Lambda(e_1 * \cdots * e_m)$ on $\Lambda^p(V)$.

18.1.3 THE STRUCTURE OF THE COMPLEX CLIFFORD ALGEBRAS.   Let $\text{Cliff}_{\mathbb{C}}(V, h) := \mathbb{C} \otimes_{\mathbb{R}} \text{Cliff}(V, h)$ be the complex Clifford algebra. Let W be a finite-dimensional complex vector space. Let c_W be a linear map from V to $\text{End}(W)$ which satisfies the *Clifford commutation relations* given in Equation (18.1.c). Then c_W extends to a unital algebra morphism from $\text{Cliff}_{\mathbb{C}}(V, h)$ to $\text{End}(W)$ which gives (W, c_W) the structure of a *unital* $\text{Cliff}_{\mathbb{C}}(V, h)$ *module*. For example, $c_\Lambda(v) := \text{ext}(v) - \text{int}(v)$ gives $\Lambda(V) \otimes_{\mathbb{R}} \mathbb{C}$ a unital $\text{Cliff}_{\mathbb{C}}(V, h)$ module structure.

Lemma 18.4

1. If $m = 2\mathsf{m}$, there exists a unique irreducible unital $\text{Cliff}_{\mathbb{C}}(V, h)$ module $\mathcal{S}_m$ of dimension 2^{m} so any unital $\text{Cliff}_{\mathbb{C}}(V, h)$ module is isomorphic to $\ell \cdot \mathcal{S}_m$.

2. If $m = 2\mathsf{m} + 1$, there exist two irreducible unital $\text{Cliff}_{\mathbb{C}}(V, h)$ modules $\mathcal{S}_m^\pm$ of dimension 2^{m} so any unital $\text{Cliff}_{\mathbb{C}}(V, h)$ module is isomorphic to $\ell_+ \cdot \mathcal{S}_m^+ \oplus \ell_- \cdot \mathcal{S}_m^-$.

3. If (W, c_W) is a unital $\text{Cliff}_{\mathbb{C}}(V, h)$ module, then there exists a Hermitian metric h_W on W so that $c_W(v)$ is skew-adjoint with respect to h_W for any $v \in V$.

4. $\text{Cliff}_{\mathbb{C}}(\mathbb{R}^{2\mathsf{m}})$ is isomorphic to the algebra $M_{2^{\mathsf{m}}}(\mathbb{C})$.

5. $\text{Cliff}_{\mathbb{C}}(\mathbb{R}^{2\mathsf{m}+1})$ is isomorphic to the algebra $M_{2^{\mathsf{m}}}(\mathbb{C}) \oplus M_{2^{\mathsf{m}}}(\mathbb{C})$.

Proof. Suppose first $m = 2\mathsf{m}$. Let $\mathcal{B} := \{e_1, f_1, e_2, f_2, \ldots, e_{\mathsf{m}}, f_{\mathsf{m}}\}$ be an orthonormal basis for V. Let $\{\tau_i := \sqrt{-1} c_W(e_i)c_W(f_i)\}_{1 \leq i \leq \mathsf{m}}$ be a commuting family of idempotents in $\text{End}(W)$.

Let Ξ be the collection of 2^{m} possible signs; $\varepsilon \in \Xi$ if $\varepsilon(i) = \pm 1$ for $1 \leq i \leq \mathrm{m}$. Let

$$E(\varepsilon) := \{w \in W : c_W(\tau_i)w = \varepsilon(i)w \text{ for } 1 \leq i \leq \mathrm{m}\}$$

be the simultaneous eigenspaces of the τ_i. We have a direct sum decomposition $W = \oplus_{\varepsilon \in \Xi} E(\varepsilon)$. Let

$$\sigma(\varepsilon) := \prod_{\nu : \varepsilon(\nu) = -1} c_W(e_\nu).$$

Since $\sigma(\varepsilon)\tau_i = \varepsilon(i)\tau_i\sigma(\varepsilon)$, $c_W(\sigma(\varepsilon))$ defines an isomorphism between $E_{\tilde\varepsilon}$ and $E_{\varepsilon\tilde\varepsilon}$ for any $\tilde\varepsilon \in \Xi$ which intertwines the action of τ_i for any i. Let $\varepsilon_0 = (+1, \ldots, +1)$ define E_0. We then have that $c_W(\sigma(\varepsilon))$ defines an isomorphism from E_0 to $E(\varepsilon)$ for any $\varepsilon \in \Xi$. Consequently, we may conclude that $\dim(W) = 2^{\mathrm{m}}\dim(E_0)$.

This shows that $\dim(W)$ is divisible by 2^{m} if W is any $\mathrm{Cliff}_{\mathbb{C}}(V, h)$ module. In particular, if $\dim(W) = 2^{\mathrm{m}}$, then necessarily W is irreducible. Let $\{\phi_\alpha\}$ be a basis for E_0 for $1 \leq \alpha \leq \ell$. Let $\phi_{\varepsilon,\alpha} := c_W(\sigma(\varepsilon))\phi_\alpha \in E(\varepsilon)$, $\mathcal{B}_\alpha := \{\phi_{\varepsilon,\alpha}\}_{\varepsilon \in \Xi}$, $W_\alpha := \mathrm{span}\{\phi_{\varepsilon,\alpha}\}_{\varepsilon \in \Xi}$. Fix i. If $\varepsilon \in \Xi$, let

$$\tilde\varepsilon = \left\{ \begin{array}{ll} \varepsilon(j) & \text{if } i \neq j \\ -\varepsilon(j) & \text{if } i = j \end{array} \right\}.$$

We then have $c_W(f_i)\phi_{\varepsilon,\alpha} = \pm\phi_{\tilde\varepsilon,\alpha}$ and $c_W(e_i)\phi_{\tilde\varepsilon,\alpha} = \mp\sqrt{-1}\phi_{\varepsilon,\alpha}$ for some canonical choice of sign $\pm$. In particular, all the W_α are isomorphic $\mathrm{Cliff}(V, h)$ modules. Since $\dim(W_\alpha) = \mathrm{m}$, they are all irreducible and Assertion 1 follows from the decomposition $W = \oplus_{1 \leq \alpha \leq \ell} W_\alpha$.

Suppose $m = 2\mathrm{m} + 1$. Let $\mathcal{B} = \{e_1, f_1, \ldots, e_{\mathrm{m}}, f_{\mathrm{m}}, g\}$ be an orthonormal basis for V. Set $\theta_{\mathcal{B}} := (\sqrt{-1})^{\mathrm{m}+1}c_W(e_1)c_W(f_1) \cdot \cdots \cdot c_W(e_{\mathrm{m}})c_W(f_{\mathrm{m}})c_W(g)$; the power of $\sqrt{-1}$ being chosen so $\theta_{\mathcal{B}}^2 = \mathrm{Id}$. If we replace g by $-g$ (i.e., if we reverse the orientation of V induced by $\mathcal{B}$), we replace $\theta_{\mathcal{B}}$ by $-\theta_{\mathcal{B}}$. Since $\theta_{\mathcal{B}}$ is central in $\mathrm{Cliff}_{\mathbb{C}}(V, h)$, $W = W_+ \oplus W_-$ decomposes as a direct sum of $\mathrm{Cliff}_{\mathbb{C}}(T^*M)$ modules into the ± 1 eigenspaces of $\theta_{\mathcal{B}}$. Thus, we can replace W by W_+ or W_-; we suppose that $W = W_+$ as the argument for W_- is analogous. Applying the argument used to establish Assertion 1 to the set $\{e_1, f_1, \ldots, e_{\mathrm{m}}, f_{\mathrm{m}}\}$, we can decompose $W = \oplus W_\alpha$ where the Clifford module structure on each W_α is standard. Since $c_W(g)$ can be recovered from the action of $c_W(e_i)$ for $i \leq m$ and $c_W(\theta_{\mathcal{B}}) = \mathrm{Id}_W$, Assertion 2 follows.

To prove Assertion 3, we note that c_Λ gives $\Lambda_{\mathbb{C}}$ a $\mathrm{Cliff}_{\mathbb{C}}(V, h)$ module structure. Furthermore, if $\|v\|^2 = 1$, then $c_\Lambda(v)$ is unitary with respect to the complexification of the inner product h_Λ defined above. Since $c_\Lambda(v)^2 = -\mathrm{Id}$, $c_\Lambda(v)$ is skew-adjoint; we rescale to see $c_\Lambda(v)$ is skew-adjoint with respect to h_Λ for any $v \in V$. If $m = 2\mathrm{m}$, then $\Lambda_{\mathbb{C}} = 2^{\mathrm{m}}\mathcal{S}_m$ so $\mathcal{S}_m$ embeds in $\Lambda_{\mathbb{C}}(V)$; we restrict h_Λ to $\mathcal{S}_m$ to establish Assertion 3 in this instance. If $m = 2\mathrm{m} + 1$, we decompose $\Lambda_{\mathbb{C}} = \ell_+\mathcal{S}_m^+ \oplus \ell_-\mathcal{S}_m^-$. If we reverse the orientation of $\mathcal{B}$, we replace $\theta_{\mathcal{B}}$ by $-\theta_{\mathcal{B}}$. Thus, $\ell_+ = \ell_- = 2^{\mathrm{m}}$ and the modules $\mathcal{S}_m^\pm$ embed in $\Lambda_{\mathbb{C}}(V)$ so again Assertion 3 follows.

Clifford multiplication defines a representation of $\mathrm{Cliff}_{\mathbb{C}}(\mathbb{R}^{2\mathrm{m}})$ in the algebra of endomorphisms of $\mathcal{S}_m$. The discussion given above shows the representation is injective. Since

$\dim(\mathrm{Cliff}_{\mathbb{C}}) = 2^m$ and $\dim(M_{\mathrm{m}}(\mathbb{C})) = (2^{\mathrm{m}})^2 = 2^m$, the representation is surjective and provides the requisite identification. The argument is analogous in the odd-dimensional setting. Assertions (4,5) follow. $\qquad\square$

We have shown that the structure of the complex Clifford algebras is periodic mod 2 and can be expressed quite easily in terms of matrix algebras. The representation theory of the real Clifford algebras is a bit more complicated as it is periodic mod 8; see Karoubi [41] for details.

Example 18.5   The modules $\mathcal{S}_m$ if $m = 2\mathrm{m}$ are defined by a collection of matrices $\{A_{m,1}, \ldots, A_{m,m}\} \subset M_{2^{\mathrm{m}}}(\mathbb{C})$ which satisfy the Clifford commutation relations given in Equation (18.1.c). To define the modules $\mathcal{S}_m^{\pm}$ if $m = 2\mathrm{m} + 1$, we again require that the matrices satisfy the Clifford commutation relations. Consequently, $A_{m,1} \cdot \cdots \cdot A_{m,m} = \sigma\,\mathrm{Id}$ where σ is a suitable 4th root of unity; the precise value of σ distinguishes $\mathcal{S}_m^+$ from $\mathcal{S}_m^-$. Let

$$A_1 := \begin{pmatrix} 0 & -1 \\ 1 & 0 \end{pmatrix}, \quad A_2 := \begin{pmatrix} 0 & \sqrt{-1} \\ \sqrt{-1} & 0 \end{pmatrix}, \quad A_3 := \begin{pmatrix} \sqrt{-1} & 0 \\ 0 & -\sqrt{-1} \end{pmatrix}.$$

We verify that $A_i A_j + A_j A_i = -2\delta_{ij}\,\mathrm{Id}$ and that $A_1 A_2 A_3 = \mathrm{Id}$. These are, of course, the relations defining the *quaternions*. We then have

1. $\pm\sqrt{-1} \in \mathbb{C}$ defines $\mathcal{S}_1^{\pm}$.
2. $\{A_1, A_2\} \subset M_2(\mathbb{C})$ defines $\mathcal{S}_2$.
3. $\{A_1, A_2, \pm A_3\} \subset M_2(\mathbb{C})$ defines $\mathcal{S}_3^{\pm}$.
4. $\{A_1 \otimes \mathrm{Id}, A_2 \otimes \mathrm{Id}, \sqrt{-1}A_3 \otimes A_1, \sqrt{-1}A_3 \otimes A_2\} \subset M_4(\mathbb{C})$ defines $\mathcal{S}_4$.
5. $\{A_1 \otimes \mathrm{Id}, A_2 \otimes \mathrm{Id}, \sqrt{-1}A_3 \otimes A_1, \sqrt{-1}A_3 \otimes A_2, \pm\sqrt{-1}A_3 \otimes A_3\} \subset M_4(\mathbb{C})$ defines $\mathcal{S}_5^{\pm}$.
6. In general, if $\{A_{m,1}, \ldots, \pm A_{m,m}\} \subset M_{2^{\mathrm{m}}}$ gives $\mathcal{S}_m^{\pm}$ for $m = 2\mathrm{m} + 1$, then
   $$\{A_1 \otimes \mathrm{Id}, A_2 \otimes \mathrm{Id}, \sqrt{-1}A_3 \otimes A_{m,1}, \ldots, \sqrt{-1}A_3 \otimes A_{m,2\mathrm{m}}\} \subset M_{2^{\mathrm{m}+1}} \text{ and}$$
   $$\{A_1 \otimes \mathrm{Id}, A_2 \otimes \mathrm{Id}, \sqrt{-1}A_3 \otimes A_{m,1}, \ldots, \pm\sqrt{-1}A_3 \otimes A_{m,2\mathrm{m}+1}\} \subset M_{2^{\mathrm{m}+1}} \text{ define}$$
   $\mathcal{S}_{m+1}$ and $\mathcal{S}_{m+2}^{\pm}$.

We caution the reader that we have not been careful in either Assertion 5 or Assertion 6b to specify exactly which matrices determine $\mathcal{S}^+$ and $\mathcal{S}^-$.

18.1.4 BUNDLES WITH CLIFFORD ALGEBRA STRUCTURES.

Let (V, h, ∇) be a geometrical vector bundle over a Riemannian manifold (M, g). We may define two-sided ideals $\mathcal{I}_{\Lambda}$ and $\mathcal{I}_{\mathrm{Cliff}}$ of the fiberwise tensor algebra of V and then use Equation (18.1.a) to define bundles $\Lambda(V)$ and $\mathrm{Cliff}(V, h)$. Lemma 18.2 then yields the extension of h to $\Lambda(V)$ and $\mathrm{Cliff}(V, h)$ is independent of any local orthonormal basis chosen. Consequently, the definitions given in Section 18.1.1 of $\Lambda(V)$, $\mathrm{Cliff}(V, h)$, ext, int, and c_{Λ} extend to this setting. We use the Levi–Civita connection $^g\nabla$ and the connection ∇_V to covariantly differentiate tensors of all types.

Let $\{s_a\}$ be a local orthonormal frame for V. Let ${}^g\Gamma_{\mu\nu}{}^\alpha$ be the Christoffel symbols of the Levi–Civita connection and let ${}^V\Gamma_{\mu a}{}^b$ be the Christoffel symbols of ∇_V. We have, for example,

$$\nabla_{\partial_{x^\mu}}\{dx^\nu \otimes s_a\} = -{}^g\Gamma_{\mu\alpha}{}^\nu dx^\alpha \otimes s_a + {}^V\Gamma_{\mu a}{}^b dx^\nu \otimes s_b,$$

$$\nabla_{\partial_{x^\mu}}(dx^{\nu_1} \wedge \cdots \wedge dx^{\nu_p})$$
$$= -\sum_\ell {}^g\Gamma_{\mu\alpha}{}^{\nu_\ell} dx^{\nu_1} \wedge \cdots \wedge dx^{\nu_{\ell-1}} \wedge dx^\alpha \wedge dx^{\nu_{\ell+1}} \wedge \cdots \wedge dx^{\nu_p}.$$

Lemma 18.6   Let (V, h, ∇) be a geometrical vector bundle over a Riemannian manifold (M, g) which is equipped with a $\mathrm{Cliff}(T^*M)$ module structure c_V.

1. If $V = T^*M$ is equipped with the canonical structures, then
$$\nabla\,\mathrm{ext}_\Lambda = 0, \quad \nabla\,\mathrm{int}_\Lambda = 0, \quad \text{and} \quad \nabla c_\Lambda = 0.$$

2. Let $\{e_1, \ldots, e_m\}$ be a local orthonormal frame for T^*M. We can choose a basis for V so that the matrix of $c_V(e_i)$ is constant relative to this basis.

3. We can choose a fiber metric on V so that $c_V(\xi)$ is skew-adjoint for $\xi \in T^*M$.

4. We can choose a unitary connection ${}^V\nabla$ on V so that ${}^V\nabla c_V = 0$.

Proof. If $V = T^*M$, then $\nabla\,\mathrm{ext}_\Lambda$, $\nabla\,\mathrm{int}_\Lambda$, and ∇c_Λ are natural; they are linear in the first derivatives of the metric with coefficients which are smooth functions of the metric tensor relative to the coordinate frames. Thus, $\nabla\,\mathrm{ext}$, $\nabla\,\mathrm{int}$, and ∇c_Λ vanish by Lemma 16.2.

   Fix a local orthonormal frame $\{e_i\}$ for T^*M. Assertion 2 is a direct consequence of the construction used to establish Lemma 18.4. Define a local fiber metric so that the local frame $\{\phi_{\varepsilon,\alpha}\}$ of Lemma 18.4 is orthonormal; that ensures that $c_V(e_i)$ is unitary and hence, since $c_V(e_i)^2 = -\,\mathrm{Id}$, skew-adjoint. A partition of unity then establishes Assertion 3; the fact that the bundles $\mathcal{S}^\pm$ need not be globally defined plays no role.

   We follow the discussion in Branson and Gilkey [11] to prove the Assertion 4. As we are dealing with a local orthonormal frame, the distinction between upper and lower indices may be suppressed. Let $c_i = c_V(e_i)$. Let Γ_{ijk} be the Christoffel symbols of the Levi–Civita connection relative to this local frame. Let ${}^V\nabla$ be defined by taking the connection 1-form $\omega_i e_i$ to be

$$\omega_i := \tfrac{1}{4}\Gamma_{ijk}c_j c_k . \tag{18.1.e}$$

This is the *spin connection* and it will play an important role subsequently in the discussion of Section 18.3.5. Let $c_{i;j}$ be the components of ∇c_V. Since the matrices c_i are constant,

$$c_{\ell;i} = \omega_i c_\ell - c_\ell \omega_i + \Gamma_{ij\ell}c_j .$$

Since the frame for the tangent bundle is orthonormal,

$$\Gamma_{ijk} = -\Gamma_{ikj} \quad \text{and} \quad c_i c_j + c_j c_i = -2\delta_{ij} .$$

We complete the proof by computing

$$c_{\ell;i} = \tfrac{1}{4}\Gamma_{ijk}(c_j c_k c_\ell - c_\ell c_j c_k) + \Gamma_{ij\ell}c_j$$
$$= \tfrac{1}{4}\Gamma_{ijk}(c_j c_k c_\ell + c_j c_\ell c_k - c_j c_\ell c_k - c_\ell c_j c_k) + \Gamma_{ij\ell}c_j$$
$$= \tfrac{1}{4}\Gamma_{ijk}(-2\delta_{k\ell}c_j + 2\delta_{j\ell}c_k) + \Gamma_{ij\ell}c_j = -\tfrac{1}{2}\Gamma_{ij\ell}c_j + \tfrac{1}{2}\Gamma_{i\ell k}c_k + \Gamma_{ij\ell}c_j$$
$$= -\Gamma_{ij\ell}c_j + \Gamma_{ij\ell}c_j = 0\,. \qquad \square$$

18.1.5 OPERATORS OF DIRAC TYPE. Let (V, h, ∇) be a geometrical vector bundle over a Riemannian manifold (M, g). The *curvature operator* is defined by setting $R_{V,ij}\phi := \phi_{;ji} - \phi_{;ij}$ for $\phi \in C^\infty(V)$. We say that an operator A on $C^\infty(V)$ is of *Dirac type* if A^2 is an operator of Laplace type. If we express $A = c_V(dx^\mu)\partial_{x^\mu} + a$, then A is of Dirac type if and only if $c_V(dx^\mu)c_V(dx^\nu) + c_V(dx^\nu)c_V(dx^\mu) = -2g^{\mu\nu}\,\mathrm{Id}$, i.e., the leading symbol of A gives a Clifford module structure to V. If E is an auxiliary endomorphism of V adopt the notation of Equation (17.3.d) to define $D_{g,\nabla,E}\phi = -(g^{\mu\nu}\phi_{;\mu\nu} + E\phi)$.

Lemma 18.7 Let (V, h, ∇) be a geometrical vector bundle over a compact Riemannian manifold (M, g). Let c_V give V a $\mathrm{Cliff}(T^*M)$ module structure. Choose (∇, h) so that $\nabla_V c_V = 0$ and $c_V(\xi)$ is skew-adjoint for $\xi \in T^*M$. Let $A = c_V \circ \nabla_V$. Then A is a self-adjoint operator of Dirac type and $A^2\phi = -\phi_{;ii} + \tfrac{1}{2}c_i c_j R_{V,ij}$ is a self-adjoint operator of Laplace type.

Proof. We sum over repeated indices. If θ is a 1-form, then $\delta\theta = -\theta_{i;i}$ by Lemma 16.4. Let $\phi, \psi \in C^\infty(V)$ and let $\theta := h(\phi, c_i\psi)e_i$. We show that A is self-adjoint by computing:

$$\begin{aligned}
0 &= -\int_M \delta\theta\,|\,\mathrm{dvol}\,| = (\phi_{;i}, c_i\psi)_{L^2} + (\phi, c_i\psi_{;i})_{L^2} + (\phi, c_{i;i}\psi)_{L^2} \\
&= -(c_i\phi_{;i}, \psi)_{L^2} + (\phi, c_i\psi_{;i})_{L^2} + 0 = (-A\phi, \psi)_{L^2} + (\phi, A\psi)_{L^2}\,.
\end{aligned}$$

It is then immediate that A^2 is self-adjoint. We complete the proof by computing that
$$A^2\phi = c_i c_j \phi_{;ji} = \tfrac{1}{2}(c_i c_j + c_j c_i)\phi_{;ji} + \tfrac{1}{2}c_i c_j(\phi_{;ji} - \phi_{;ij}) = -\phi_{;ii} + \tfrac{1}{2}c_i c_j R_{ij}\phi\,. \quad \square$$

18.2 THE DE RHAM COMPLEX

In Section 18.2.1, we specialize the notions of Section 18.1 to the bundle of exterior differential forms. We show that the Laplace–Beltrami operator is an operator of Laplace type. In Section 18.2.2, we establish the Weitzenböck formula. In Section 18.2.3, we discuss the Hodge–de Rham Theorem. In Section 18.2.4, we discuss the Chern–Gauss–Bonnet Theorem. In Section 18.2.5, we use the Weitzenböck formula to prove the Bochner Vanishing Theorem. In Section 18.2.6, we treat Poincaré Duality.

18.2.1 THE HODGE–LAPLACE–BELTRAMI OPERATOR. To simplify the notation, we set $\Lambda M = \oplus_p \Lambda^p M$ where $\Lambda^p M = \Lambda^p(T^*M)$. Use Lemma 16.4 to express exterior differentiation d and the adjoint δ in the form:

$$d = \mathrm{ext}(e_i)\nabla_{e_i}\,, \qquad \delta = -\mathrm{int}(e_i)\nabla_{e_i}\,, \qquad d + \delta = c_\Lambda(e_i)\nabla_{e_i}\,.$$

We use Lemmas 18.6 and 18.7 to see that $d + \delta$ is a self-adjoint operator of Dirac type on $C^\infty(\Lambda M)$ and the *Hodge–Laplace–Beltrami operator* $\Delta = (d + \delta)^2$ is of an operator of Laplace type. Since $d^2 = 0$, we have dually that $\delta^2 = 0$. Consequently, $\Delta = d\delta + \delta d$ respects the grading of ΛM and we may decompose $\Delta = \oplus_p \Delta_p$ where

$$\Delta_p := d_{p-1}\delta_p + \delta_{p+1}d_p$$

is a self-adjoint operator of Laplace type on $C^\infty(\Lambda^p M)$. Let R_{ij} denote the curvature operator of the Levi–Civita connection acting on the exterior algebra:

$$R_{ij}(e_{k_1} \wedge \cdots \wedge e_{k_p}) = \sum_{\nu=1}^{p}(-1)^{\nu-1} e_{k_1} \wedge \cdots \wedge e_{k_{\nu-1}} \wedge \left\{ \sum_{\ell=1}^{m} R_{ijk_\nu \ell} e_\ell \right\} \wedge e_{k_{\nu+1}} \wedge \cdots \wedge e_{k_p}.$$

18.2.2 THE WEITZENBÖCK FORMULA.

Lemma 18.8   Let (M, g) be a compact Riemannian manifold, let $\{e_i\}$ be a local orthonormal frame for TM, let $c_\Lambda(\xi) = \text{ext}(\xi) - \text{int}(\xi)$, let $\rho(e_i) = R_{kijk}e_j$ be the *Ricci operator*, and let $\theta \in C^\infty(\Lambda M)$.

1. $\Delta\theta = -\theta_{;ii} + \frac{1}{2}c_\Lambda(e_i)c_\Lambda(e_j)R_{ij}\theta$.
2. If $p = 1$, then $(\Delta\theta, \theta)_{L^2} = (\nabla\theta, \nabla\theta)_{L^2} + (\rho\theta, \theta)_{L^2}$.

Proof. We specialize Lemma 18.7 to the setting at hand to obtain Assertion 1. If we set $p = 1$, we obtain

$$(\Delta\theta, \theta)_{L^2} = (-\theta_{;ii}, \theta)_{L^2} + (\tfrac{1}{2}c_\Lambda(e_i)c_\Lambda(e_j)R_{ij}\theta, \theta)_{L^2}.$$

We use Lemma 16.4 to see $(-\theta_{;ii}, \theta)_{L^2} = (\theta_{;i}, \theta_{;i})_{L^2}$. We may express

$$\tfrac{1}{2}c_\Lambda(e_i)c_\Lambda(e_j)R_{ij}\theta = \tfrac{1}{2}\theta_k c_\Lambda(e_i)c_\Lambda(e_j)R_{ijk\ell}e_\ell = \theta_k R_{ijk\ell}c_\Lambda(e_i)c_\Lambda(e_j)c_\Lambda(e_\ell)\mathbb{1}. \quad (18.2.a)$$

Since $R_{ijk\ell} = -R_{jik\ell}$ we may sum over $i < j$. If $\ell \neq i$ and $\ell \neq j$, we can cyclically permute the factors. The first Bianchi identity $R_{ijk\ell} + R_{jki\ell} + R_{kij\ell} = 0$ then shows that Equation (18.2.a) is zero. Consequently, either $i = \ell$ or $j = \ell$ so

$$\tfrac{1}{2}c_\Lambda(e_i)c_\Lambda(e_j)R_{ij}\theta = -R_{i\ell k\ell}\theta_k e_i = \rho_{ik}\theta_k e_i. \qquad \square$$

18.2.3 THE HODGE–DE RHAM THEOREM. Because $d^2 = 0$,

$$\mathcal{E}_{\text{deR}} := \{d_p : C^\infty(\Lambda^p M) \to C^\infty(\Lambda^{p+1} M)\}_{0 \leq p \leq m-1}$$

is an *elliptic complex of Dirac type*. This elliptic complex is called the *de Rham complex*, and the analysis of Section 17.4.5 applies. Following the discussion in Section 17.4.5, we define the de Rham cohomology groups to be

$$H_{\text{deR}}^p(M) := \frac{\ker\{d : C^\infty(\Lambda^p M) \to C^\infty(\Lambda^{p+1} M)\}}{\text{Image}\{d : C^\infty(\Lambda^{p-1} M) \to C^\infty(\Lambda^p M)\}}.$$

We refer to Section 5.1 of Book II for further details. Assertion 1 in the following result identifies H_{deR}^p with the topological cohomology groups $H^p(M, \mathbb{C})$. It is due to de Rham [19] and we refer to the discussion in Section 5.1 of Book II for the proof and for further details. Assertion 2 in the following result is due to Hodge [39] and follows from Hodge Decomposition Theorem (see Theorem 17.24).

Theorem 18.9   Let $\Delta_p := (\delta_{p+1}d_p + d_{p-1}\delta_p)$ be the Laplacian on p-forms over a compact Riemannian manifold (M, g).

1. There is a natural isomorphism between $H_{\mathrm{deR}}^p(M)$ and $H^p(M, \mathbb{C})$.

2. The natural inclusion of $\ker\{\Delta_p\}$ into $H_{\mathrm{deR}}^p(M)$ identifies $H_{\mathrm{deR}}^p(M)$ with $\ker\{\Delta_p\}$.

3. Let $d\theta = 0$. Then $[\theta] = 0$ in $H_{\mathrm{deR}}^p(M)$ if and only if $\theta \in \mathrm{range}\{\Delta_p\}$.

18.2.4 THE CHERN–GAUSS–BONNET THEOREM. Let M be compact. Theorem 18.9 shows that

$$\mathrm{Index}(\mathcal{E}_{\mathrm{deR}}) = \sum_p (-1)^p \dim(H^p(M, \mathbb{R})) = \chi(M)$$

where $\chi(M)$ is the Euler–Poincaré characteristic of M. Let E_m be the *Euler form* (see Section 16.1.6.g). Chern [16] generalized the Gauss–Bonnet Theorem (see Theorem 3.23 of Book I) proving:

Theorem 18.10   If (M, g) is a compact Riemannian manifold, then

$$\mathrm{Index}(\mathcal{E}_{\mathrm{deR}}) = \int_M E_m |\mathrm{dvol}|.$$

18.2.5 THE BOCHNER VANISHING THEOREM [10].

Theorem 18.11   Let (M, g) be a compact, connected Riemannian manifold. Assume the Ricci curvature ρ is non-negative and positive at some point. Then $H_{\mathrm{deR}}^1(M) = 0$.

Proof. By Lemma 18.8, if f is a smooth one-form, then

$$(\Delta f, f)_{L^2} = (\nabla f, \nabla f)_{L^2} + (\rho f, f)_{L^2} \geq (\rho f, f)_{L^2}.$$

If θ is a harmonic 1-form, then $0 \geq (\rho\theta, \theta)_{L^2}$; as ρ is non-negative, this implies $\rho\theta = 0$ and we have equality. Thus, $(\nabla\theta, \nabla\theta)_{L^2} = 0$ so θ is parallel. Since ρ is positive at some point, we have $\theta(x_0) = 0$ for some point. Since $\nabla\theta = 0$, this implies $\theta \equiv 0$. Thus, $\ker\{\Delta_1\} = 0$. We use Theorem 18.9 to identify $\ker\{\Delta_1\}$ with $H_{\mathrm{deR}}^1(M)$.   □

If we had assumed the Ricci curvature is positive definite at every point, it would follow that the fundamental group of M is finite. Since $H_{\text{deR}}^1(M) \approx \text{Hom}(\pi_1(M), \mathbb{R})$ the theorem would follow in this special case.

18.2.6 POINCARÉ DUALITY [59]. We now derive material we will need subsequently in Section 18.5.3 when we discuss Serre duality; a different treatment was given in Book II (see Theorem 5.14) based on the Hodge $\star$ operator. Let (M, g) be a compact, oriented Riemannian manifold of dimension m. Let $\{e_1, \ldots, e_m\}$ be an oriented orthonormal basis for the tangent bundle TM. The orientation form is then given by $\text{orn} = e_1 \wedge \cdots \wedge e_m$. Clifford multiplication by the orientation form $c_\Lambda(\text{orn}) = c_\Lambda(e_1) \cdot \cdots \cdot c_\Lambda(e_m)$ (see Section 18.1.3) is then an invariantly defined isometry of ΛM which satisfies $c_\Lambda(\text{orn}) = (-1)^{m(m-1)/2} \text{Id}$. It maps $\Lambda^p M$ to $\Lambda^{m-p}(M)$ and is equivalent to the Hodge $\star$ operator modulo sign by Lemma 18.3.

Theorem 18.12   Let (M, g) be a compact, oriented Riemannian manifold of dimension m.

1. $(d + \delta)c_\Lambda(\text{orn}) = (-1)^{m-1}c_\Lambda(\text{orn})(d + \delta)$.
2. $c_\Lambda(\text{orn})$ induces an isomorphism from $H_{\text{deR}}^p(M)$ to $H_{\text{deR}}^{m-p}(M)$.

Proof. By Lemma 16.4, $d = \text{ext} \circ \nabla$ and $\delta = -\text{int} \circ \nabla$ so $d + \delta = c_\Lambda \circ \nabla$ where c_Λ denotes Clifford multiplication. Thus, $c_\Lambda(\text{orn})(d + \delta) = c_\Lambda(\text{orn}) \circ c_\Lambda \circ \nabla$. If ξ is a cotangent vector, then we can choose the orthonormal basis so ξ is a multiple of e_1. Since

$$
\begin{aligned}
c_\Lambda(e_1)c_\Lambda(\text{orn}) &= c_\Lambda(e_1 * e_1 * e_2 * \cdots * e_m) = (-1)^{m-1}c_\Lambda(e_1 * e_2 * \cdots * e_m * e_1) \\
&= (-1)^{m-1}c_\Lambda(\text{orn})c_\Lambda(e_1),
\end{aligned}
$$

we conclude $c_\Lambda(\text{orn}) \circ c_\Lambda \circ \nabla = (-1)^{(m-1)}c_\Lambda \circ (\text{Id} \otimes c_\Lambda(\text{orn})) \circ \nabla$. It is immediate from the definition that $\mathcal{E} := (\text{Id} \otimes c_\Lambda(\text{orn})) \circ \nabla - \nabla \circ c_\Lambda(\text{orn})$ is linear in the one-jets of the metric. Since this operator is natural and since we can always choose the coordinate system in question so the first derivatives of the metric vanish at the origin, $\mathcal{E} = 0$. We establish Assertion 1 by computing $c_\Lambda \circ (\text{Id} \otimes c_\Lambda(\text{orn})) \circ \nabla = (-1)^{(m-1)}c_\Lambda \circ \nabla \circ c_\Lambda(\text{orn})$. Assertion 2 follows from the isomorphism of Theorem 18.9.   □

18.3   SPINORS

In Sections 18.3.1and 18.3.2, we define the orthogonal and spin groups and establish some of their basic properties. We discussed the Stiefel–Whitney classes quite generally in Section 16.1.6. In Section 18.3.3, we discuss the first Stiefel–Whitney class in some detail and show that if V is a real vector bundle, then $\text{sw}_1(V)$ vanishes if and only if V is orientable. In Section 18.3.4, we discuss the second Stiefel–Whitney class and show that if V is orientable, then $\text{sw}_2(V) = 0$ if and only if V admits a spin structure. In Section 18.3.5, we define the spinor bundle, define the spinor connection, define the spin Dirac operator, and obtain the Lichnerowicz formula. In Section 18.3.6, we use the Lichnerowicz formula to construct obstructions to the existence of metrics of positive scalar curvature.

18.3.1 THE ORTHOGONAL GROUP.
Let h be a positive definite inner product on a real vector space V of dimension m. Let

$$O(V, h) := \{\phi \in \mathrm{GL}(V) : \|\phi v\|^2 = \|v\|^2 \ \forall v \in V\}$$

be the *orthogonal group*. The *polarization identity* $(\xi, \eta) = \frac{1}{4}\{\|\eta + \xi\|^2 - \|\eta - \xi\|^2\}$ expresses the inner product in terms of the norm and shows, equivalently,

$$O(V, h) = \{\phi \in \mathrm{GL}(V) : \phi^* h = h\}\,.$$

Let $O(\ell) = O(\mathbb{R}^m, h_e)$ where h_e is the standard Euclidean inner product on $\mathbb{R}^m$. We may use an orthonormal basis for V to identify $O(V, h)$ with $O(\ell)$ and obtain

$$O(V, h) = \{\phi \in M_m(\mathbb{R}) : \phi \cdot \phi^t = \mathrm{Id}\}\,.$$

This implies that $\det(\phi)^2 = \det(\phi)\det(\phi^t) = \det(\phi\phi^t) = \det(\mathrm{Id}) = 1$ so $\det(\phi) = \pm 1$. The *special orthogonal group* is defined by setting

$$\mathrm{SO}(V, h) := \{\phi \in O(V, h) : \det(\phi) = 1\} \subset O(V, h)\,;$$

see the discussion in Section 6.4 of Book II for further details. Let $\mathrm{SO}(\ell) = \mathrm{SO}(\mathbb{R}^m, h_e)$. The groups $O(V, h)$ and $\mathrm{SO}(V, h)$ are closed subgroups of the Lie group $\mathrm{GL}(V)$ and, consequently, by Theorem 6.10 of Book II are Lie groups. We identify S^1 with the unit complex numbers and S^3 with the unit quaternions to give S^1 and S^3 Lie group structures.

Lemma 18.13

1. $\mathrm{SO}(2) = S^1$ and $\pi_1(\mathrm{SO}(2)) = \mathbb{Z}$.
2. $\mathrm{SO}(3) = S^3/\mathbb{Z}_2 = \mathbb{RP}^3$ and $\pi_1(\mathrm{SO}(3)) = \mathbb{Z}_2$.
3. $\mathrm{SO}(4) = (S^3 \times S^3)/\mathbb{Z}^2$ and $\pi_1(\mathrm{SO}(4)) = \mathbb{Z}_2$.
4. $\pi_1(\mathrm{SO}(k)) = \mathbb{Z}_2$ for $k \geq 5$.

Proof. Assertions 1–3 follow from Lemma 6.24 of Book II. In perhaps an excess of zeal, we sketch the proof of Assertion 4 at the urging of our colleague and friend E. Puffini. It uses algebraic topology. Let $\pi_k(M)$ denote the kth homotopy group of M; $\pi_1(M)$ is the fundamental group. Let S be the south pole of S^m. The map $\phi \to \phi \cdot S$ defines a principal bundle

$$\mathrm{SO}(m) \to \mathrm{SO}(m+1) \to S^m\,;$$

we refer to Theorem 7.1 of Book II for further details. We suppose that $m \geq 3$ so that the first two homotopy groups $\pi_1(S^m)$ and $\pi_2(S^m)$ vanish. The *long exact sequence of a fibration* then establishes Assertion 4 by induction since it gives rise to a short exact sequence

$$0 = \pi_2(S^m) \to \pi_1(\mathrm{SO}(m)) \to \pi_1(\mathrm{SO}(m+1)) \to \pi_1(S^m) = 0. \qquad \square$$

18.3.2 THE SPIN GROUP. Let $S(V, h) := \{v \in V : h(v, v) = 1\}$ be the unit sphere in V. Let

$$\text{Spin}(V, h) := \{\xi := v_1 * \cdots * v_{2p} : v_i \in S(V, h)\} \subset \text{Cliff}^+(V, h)$$

be the collection of all products of an even number of unit vectors in V with the subspace topology. If $\xi \in \text{Spin}(V, h)$, then $\xi * \xi^t = 1$ so $S(V, h)$ is a topological group. If $v \in V$, let $\varrho(\xi)v = \xi * v * \xi^{-1}$. Let $\{e_1, \ldots, e_m\}$ be an orthonormal basis for (V, h);

$$e_1 * e_i * e_1 = \left\{ \begin{array}{ll} -e_1 & \text{if } i = 1 \\ e_i & \text{if } i > 1 \end{array} \right\}.$$

This is reflection in the hyperplane defined by e_1. Consequently, $\varrho(v_1 * \cdots * v_{2p})$ is the product of an even number of hyperplane reflections and hence belongs to $\text{SO}(V, h)$. Since

$$\varrho(\xi_1 * \xi_2) = \varrho(\xi_1)\varrho(\xi_2),$$

ϱ defines a group homomorphism from $\text{Spin}(V, h)$ to $\text{SO}(V, h)$. If $\{f_1, f_2\}$ is an orthonormal set and if $\theta \in \mathbb{R}$, set

$$\begin{aligned}
\xi_{\theta, f_1, f_2} &:= \{\cos(\tfrac{1}{4}\theta) f_1 + \sin(\tfrac{1}{4}\theta) f_2\} * \{-\cos(\tfrac{1}{4}\theta) f_1 + \sin(\tfrac{1}{4}\theta) f_2\} \\
&= \{\cos^2(\tfrac{1}{4}\theta) - \sin^2(\tfrac{1}{4}\theta)\} \, 1 + 2\sin(\tfrac{1}{4}\theta) \cos(\tfrac{1}{4}\theta) f_1 * f_2 \qquad (18.3.a) \\
&= \cos(\tfrac{1}{2}\theta) 1 + \sin(\tfrac{1}{2}\theta) f_1 * f_2 \in \text{Spin}(V, h).
\end{aligned}$$

Let R_{θ, f_1, f_2} be the rotation through the angle θ in the plane spanned by $\{f_1, f_2\}$, i.e.,

$$R_{\theta, f_1, f_2} v = \left\{ \begin{array}{ll} \cos(\theta) f_1 + \sin(\theta) f_2 & \text{if } v = f_1 \\ -\sin(\theta) f_1 + \cos(\theta) f_2 & \text{if } v = f_2 \\ v & \text{if } v \perp \text{span}\{f_1, f_2\} \end{array} \right\} \in \text{SO}(V, h). \qquad (18.3.b)$$

Lemma 18.14 We have that $\varrho(\xi_{\theta, f_1, f_2}) = R_{\theta, f_1, f_2}$.

Proof. We compute:

$$\begin{aligned}
\varrho(\xi_{\theta, f_1, f_2}) f_1 &= \{\cos(\tfrac{1}{2}\theta) 1 + \sin(\tfrac{1}{2}\theta) f_1 * f_2\} * f_1 * \{\cos(\tfrac{1}{2}\theta) 1 + \sin(\tfrac{1}{2}\theta) f_2 * f_1\} \\
&= \cos^2(\tfrac{1}{2}\theta) f_1 + \cos(\tfrac{1}{2}\theta) \sin(\tfrac{1}{2}\theta) f_2 + \sin(\tfrac{1}{2}\theta) \cos(\tfrac{1}{2}\theta) f_2 - \sin^2(\tfrac{1}{2}\theta) f_1 \\
&= \cos(\theta) f_1 + \sin(\theta) f_2, \\
\varrho(\xi_{\theta, f_1, f_2}) f_2 &= \{\cos(\tfrac{1}{2}\theta) 1 + \sin(\tfrac{1}{2}\theta) f_1 * f_2\} * f_2 * \{\cos(\tfrac{1}{2}\theta) 1 + \sin(\tfrac{1}{2}\theta) f_2 * f_1\} \\
&= \cos^2(\tfrac{1}{2}\theta) f_2 - \cos(\tfrac{1}{2}\theta) \sin(\tfrac{1}{2}\theta) f_1 - \sin(\tfrac{1}{2}\theta) \cos(\tfrac{1}{2}\theta) f_1 - \sin^2(\tfrac{1}{2}\theta) f_2 \\
&= -\sin(\theta) f_1 + \cos(\theta) f_2, \\
\varrho(\xi_{\theta, f_1, f_2}) v &= \{\cos(\tfrac{1}{2}\theta) 1 + \sin(\tfrac{1}{2}\theta) f_1 * f_2\} * v * \{\cos(\tfrac{1}{2}\theta) 1 + \sin(\tfrac{1}{2}\theta) f_2 * f_1\} \\
&= v * \{\cos(\tfrac{1}{2}\theta) 1 + \sin(\tfrac{1}{2}\theta) f_1 * f_2\} * \{\cos(\tfrac{1}{2}\theta) 1 + \sin(\tfrac{1}{2}\theta) f_2 * f_1\} \\
&= v \quad \text{if} \quad v \perp \text{span}\{f_1, f_2\}. \qquad \square
\end{aligned}$$

Lemma 18.15   Let $m \geq 2$.

1. ϱ is a surjective group homomorphism from $\mathrm{Spin}(V, h)$ to $\mathrm{SO}(V, h)$.
2. $\ker\{\varrho\} = \{\pm \mathbb{1}\}$.
3. $\mathrm{Spin}(V, h)$ is an arc-connected compact Lie group.
4. $\varrho : \mathrm{Spin}(V, h) \to \mathrm{SO}(V, h)$ is a covering projection.
5. If $m > 2$, then $\mathrm{Spin}(V, h)$ is the universal cover of $\mathrm{SO}(V, h)$.

Proof. We already showed that $\varrho : \mathrm{Spin}(V, h) \to \mathrm{SO}(V, h)$ is a group homomorphism. Let $\dim(V) = m$ where $m = 2\mathfrak{m}$ or $m = 2\mathfrak{m} + 1$. Let $\phi \in \mathrm{SO}(V, h)$. The complex eigenvalues of ϕ all have unit length and appear in conjugate pairs; if λ is an eigenvalue of ϕ, then $|\lambda| = 1$ and $\bar\lambda$ is also an eigenvalue of ϕ. Furthermore, 1 is an eigenvalue of ϕ if m is odd. This shows that there is an orthonormal basis $\{e_1, \dots, e_m\}$ for (V, h) and angles θ_i so that

$$\phi = R_{\theta_1, e_1, e_2} R_{\theta_2, e_3, e_4} \cdots R_{\theta_\mathfrak{m}, e_{2\mathfrak{m}-1}, e_{2\mathfrak{m}}} \tag{18.3.c}$$
$$= \varrho(\xi_{\theta_1, e_1, e_2} * \cdots * \xi_{\theta_\mathfrak{m}, e_{2\mathfrak{m}-1}, e_{2\mathfrak{m}}}) \, .$$

This shows that ϱ is surjective which proves Assertion 1.

We use Lemma 18.2 to establish Assertion 2. If $\varrho(\xi) = \mathrm{Id}$, then $\xi * v * \xi^{-1} = v$ for all v in V. Consequently, ξ belongs to the center of $\mathrm{Cliff}(V, h)$. If m is even, then $\xi = a\mathbb{1}$ is a scalar. Since $\mathrm{Id} = \xi * \xi^t$, we have $a^2 = 1$ so $a = \pm 1$ and Assertion 2 follows. If m is odd, then

$$\xi = a_0 \mathbb{1} + a_1 e_1 * \cdots * e_m \, .$$

Since $\xi \in \mathrm{Cliff}^+(V, h)$ and $e_1 * \cdots * e_m \in \mathrm{Cliff}^-(V, h)$, we conclude $a_1 = 0$. The same argument implies $a_0^2 = 1$ and completes the proof of Assertion 2.

If $\xi \in \mathrm{Spin}(V, h)$, then we set $\phi := \varrho(\xi) \in \mathrm{SO}(V, h)$. By Equations (18.3.a) and (18.3.c), there exists unit vectors $\{v_1, \dots, v_{2\mathfrak{m}}\}$ in V so $\varrho(\xi) = \varrho(v_1 * \cdots * v_{2\mathfrak{m}})$. By Assertion 2,

$$\xi = \pm v_1 * v_2 * \cdots * v_{2\mathfrak{m}} \, .$$

By replacing v_1 by $-v_1$ if necessary, we can assume $\xi = v_1 * v_2 \cdots * v_{2\mathfrak{m}}$. Consequently, Clifford multiplication defines a natural map from the $2\mathfrak{m}$-fold Cartesian product of $S(V, h)$ onto $\mathrm{Spin}(V, h)$. Since $m \geq 2$, $S(V, h)$ is compact and connected; consequently, $\mathrm{Spin}(V, h)$ is compact and connected as well. Left Clifford multiplication embeds $\mathrm{Spin}(V, h)$ as a compact subset of $\mathrm{GL}(\mathrm{Cliff}(V, h))$. A closed subgroup of a Lie group is a Lie group by Theorem 6.10 of Book II. Consequently, $\mathrm{Spin}(V, h)$ is a compact connected Lie group. This proves Assertion 3.

We may express $\varrho(\xi) = L_\xi R_{\xi^t}$ as the composition of left and right Clifford multiplication. Consequently, ϱ is smooth. Let $\mathfrak{spin}$ and $\mathfrak{so}$ be the Lie algebras of $\mathrm{Spin}(V, h)$ and $\mathrm{SO}(V, h)$. Since ϱ is locally 1-1, ϱ_* identifies $\mathfrak{spin}$ with $\mathfrak{so}$ and shows ϱ is a covering projection. If $m > 2$, then $\pi_1(\mathrm{SO}(V, h)) = \mathbb{Z}_2$ by Lemma 18.13. Since ϱ is 2-to-1 and $\mathrm{Spin}(V, h)$ is connected, it follows that $\mathrm{Spin}(V, h)$ is the universal cover of $\mathrm{SO}(V, h)$. $\square$

We now examine the relationship between the Lie algebra $\mathfrak{spin}(V, h)$ of $\mathrm{Spin}(V, h)$ and the Lie algebra $\mathfrak{so}(V, h)$ of $\mathrm{SO}(V, h)$. Let $\{f_1, f_2\}$ be an orthonormal set. Let

$$\mathfrak{so}_{f_1, f_2} := \partial_\theta R_{\theta, f_1, f_2}|_{\theta=0} : v \to \left\{ \begin{array}{ll} f_2 & \text{if } v = f_1 \\ -f_1 & \text{if } v = f_2 \\ 0 & \text{if } v \perp \mathrm{span}\{f_1, f_2\} \end{array} \right\}. \tag{18.3.d}$$

Lemma 18.16   Let $\{e_1, \ldots, e_m\}$ be an orthonormal basis for (V, h). Set $\mathfrak{so}_{ij} := \mathfrak{so}_{e_i, e_j}$ and $\mathfrak{sp}_{ij} := \frac{1}{2} e_i * e_j$. Then $\{\mathfrak{so}_{ij}\}_{1 \leq i < j \leq m}$ is a basis for $\mathfrak{so}(V, h)$, $\{\mathfrak{sp}_{ij}\}_{1 \leq i < j \leq m}$ is a basis for $\mathfrak{spin}(V, h)$, and $\varrho_* : \frac{1}{2} e_i * e_j \to \mathfrak{so}_{ij}$.

Proof. Equation (18.3.b) shows that the map $\theta \to R_{\theta, e_i, e_j}$ defines a smooth one-parameter subgroup of $\mathrm{SO}(V, h)$; consequently, differentiating at the origin yields an element of $\mathfrak{so}(V, h)$. The Lie algebra of $\mathrm{SO}(V, h)$ consists of the skew-symmetric endomorphisms (see Section 6.4 of Book II). Consequently, $\{\mathfrak{so}_{ij}\}_{1 \leq i < j \leq m}$ is basis for $\mathfrak{so}(V, h)$. This proves Assertion 1. Since ϱ is a covering projection, it is a group isomorphism near the identity. Since $\varrho(\xi_{\theta, e_i, e_j}) = R_{\theta, e_i, e_j}$, the map $\theta \to \xi_{\theta, e_i, e_j}$ is a one-parameter subgroup of $\mathrm{Spin}(V, h)$ and $\{\frac{1}{2} e_i * e_j\}_{1 \leq i < j \leq m}$ is a basis for $\mathfrak{spin}(V, h)$. $\qquad \square$

18.3.3 THE FIRST STIEFEL–WHITNEY CLASS.

Let (M, g) be a compact connected Riemannian manifold. Let $\mathcal{U} := \{\mathcal{O}_i\}_{i=1}^n$ be a finite open cover M by small geodesic balls. Since sufficiently small geodesic balls are geodesically convex, $\mathcal{U}$ forms a simple cover of M since the finite intersection of geodesically convex sets is either empty or geodesically convex and hence contractible. We consider a set of indices $I = \{1 \leq i_0 < \cdots < i_p \leq n\}$. Set

$$\mathcal{O}_I = \mathcal{O}_{i_0} \cap \cdots \cap \mathcal{O}_{i_p}.$$

We introduced sheaf cohomology in Section 16.1.5. We recall the definition briefly as follows. Let $\mathbb{Z}_2$ be the group with two elements $\{\pm 1\}$. An element Φ^p of $C^p(M, \mathbb{Z}_2)$ is an assignment $I \to \Phi^p(I) \in \mathbb{Z}_2$ for $|I| = p + 1$ where we set $\Phi^p(I) = 1$ if $\mathcal{O}_I$ is empty. Let J be a collection of indices $\{1 \leq j_0 < \cdots < j_{p+1} \leq n\}$. Since we are using multiplicative notation, the co-boundary $\delta(\Phi^p) \in C^{p+1}(\mathbb{Z}_2)$ is defined multiplicatively:

$$\delta(\Phi^p)(J) = \prod_{\nu=0}^p \Phi^p(j_0, \ldots, j_{\nu-1}, j_{\nu+1}, \ldots, j_{p+1}) \in \mathbb{Z}_2.$$

One has $\delta^2 = 1$ and the cohomology groups

$$H^p(M, \mathbb{Z}_2) := \frac{\ker\{\delta : C^p(M, \mathbb{Z}_2) \to C^{p+1}(M, \mathbb{Z}_2)\}}{\mathrm{range}\{\delta : C^{p-1}(M, \mathbb{Z}_2) \to C^p(M, \mathbb{Z}_2)\}}$$

are independent of the particular simple cover which was chosen.

Let (V, h) be a real vector bundle over M which is equipped with a positive definite fiber metric h. Let $S = \{\vec{s}_i\}$ be a collection of local orthonormal frames for V over the $\mathcal{O}_i$. Over $\mathcal{O}_i \cap \mathcal{O}_j$, we may express $\vec{s}_i = \phi_S(i, j) \cdot \vec{s}_j$ where the transition functions

$$\phi_S(i, j) : \mathcal{O}_i \cap \mathcal{O}_j \to \mathrm{O}(\ell) := \mathrm{O}(\mathbb{R}^\ell).$$

Define an element of $\mathcal{C}^1(M, \mathbb{Z}_2)$ by setting $\Phi_S^1(i, j) = \det(\phi_S(i, j)) = \pm 1$. We observe that

$$\phi_S(i, j)\phi_S(j, k) = \phi_S(i, k) \quad \text{on} \quad \mathcal{O}_i \cap \mathcal{O}_j \cap \mathcal{O}_k.$$

We show that $\delta(\Phi_S^1) = 0$ by computing:

$$\begin{aligned}
\delta(\Phi_S^1)(i, j, k) &= \Phi_S^1(i, j)\Phi_S^1(j, k)\Phi_S^1(i, k) = \det\{\phi_S(i, j)\}\det\{\phi_S(j, k)\}\det\{\phi_S(i, k)\} \\
&= \det\{\phi_S(i, j)\phi_S(j, k)\phi_S(k, i)\} = +1.
\end{aligned}$$

Consequently, $[\Phi_S^1]$ determines an element of $H^1(M, \mathbb{Z}_2)$. Suppose we had chosen a different collection $T = \{\vec{t}_i\}$ of local frames for V over each $\mathcal{O}_i$. We then have $\vec{t}_i = \phi_{TS}(i)\vec{s}_i$ where $\phi_{TS}(i) : \mathcal{O}_i \to \mathrm{O}(\ell)$. Thus, $\phi_T(i, j) = \phi_{TS}(i)\phi_S(i, j)\phi_{TS}(j)$. Define $\Phi_{TS}^0 \in \mathcal{C}^0(M, \mathbb{Z}_2)$ by setting $\Phi_{TS}^0(i) := \det\{\phi_{TS}(i)\}$. Then $\Phi_T^1 = \delta\Psi_{TS}^0 \cdot \Phi_S^1$ since $\Phi_T^1(i, j) = \Phi_{TS}^0(i)\Psi_{TS}^0(j)\Phi_S^1(i, j)$. This shows that $[\Phi_S^1] = [\Phi_T^1]$ in $H^1(M, \mathbb{Z}_2)$ is independent of the particular choice of frames. We denote this class by $\mathrm{sw}_1(V) \in H^2(M, \mathbb{Z}_2)$; it is independent of the particular fiber metric h on V which is chosen.

Lemma 18.17 Let V be a real vector bundle over a compact manifold M. Then V is orientable if and only if $\mathrm{sw}_1(V)$ is trivial.

Proof. If V is orientable, we can choose local frames so $\det(\phi_S(i, j)) = 1$ which implies $\mathrm{sw}_1(V)$ is trivial. Conversely, if $\mathrm{sw}_1(V)$ is trivial, then

$$\Phi_S^1 = \delta(\Phi^0) \quad \text{for some} \quad \Phi^0 \in \mathcal{C}^0(M, \mathbb{Z}_2).$$

If $\Phi^0(i) = +1$, we do not change $\vec{s}_i$, while if $\Phi^0(i) = -1$, we change the orientation of $\vec{s}_i$. This creates a collection of local frames where $\det(\phi_S(i, j)) = +1$ and hence V is orientable. □

Let $\mathrm{Vect}_\mathbb{R}^1(M)$ be the set of isomorphism classes of real line bundles over M. Tensor product makes $\mathrm{Vect}_\mathbb{R}^1(M)$ into an Abelian group. If (V, h) is a real line bundle, then there is a unique metric compatible connection ∇_V on V. The curvature of this connection is trivial so the holonomy defined by parallel translation gives rise to a representation $\sigma_L : \pi_1(M) \to \mathbb{Z}_2$.

Lemma 18.18

1. $\mathrm{sw}_1 : \mathrm{Vect}_\mathbb{R}^1(M) \to H^1(M, \mathbb{Z}_2)$ is a group isomorphism.
2. $\sigma_L : \mathrm{Vect}_\mathbb{R}^1(M) \to \mathrm{Hom}(\pi_1(M); \mathbb{Z}_2)$ is a group isomorphism.

Proof. Let $\{s_i\}$ be local unit length sections to a line bundle L_1 and let $\{t_i\}$ be local unit length sections to a line bundle L_2 over M. Then $\{s_i \otimes t_i\}$ are unit length local sections to the line bundle $L_1 \otimes L_2$. Let $s_i = \varepsilon_1(i,j)s_j$ and $t_i = \varepsilon_2(i,j)t_j$. We then have $s_i t_i = \varepsilon_1(i,j)\varepsilon_2(i,j)s_j t_j$. From this it follows that $\mathrm{sw}_1(L_1 \otimes L_2)$ is the product of $\mathrm{sw}_1(L_1)$ and $\mathrm{sw}_2(L_2)$ in $H^1(M, \mathbb{Z}_2)$ and hence sw_1 is a group homomorphism. Given $\varepsilon \in \mathcal{C}^1(M, \mathbb{Z}_2)$, we can define gluing functions $s_i = \varepsilon(i,j)s_j$; the cocycle condition ensures these glue together properly to define a real line bundle which is exactly the condition that $\delta(\varepsilon)$ is trivial. Thus, sw_1 is surjective. Suppose that $\mathrm{sw}_1(L_1) = \mathrm{sw}_1(L_2)$. This implies $\mathrm{sw}_1(L_1 \otimes L_2)$ is trivial and hence $L_1 \otimes L_2$ is orientable by Lemma 18.17. A real line bundle is trivial if and only if it is orientable and, consequently, $L_1 \otimes L_2$ is trivial. This implies L_1 is isomorphic to L_2 and hence sw_1 is injective. This proves Assertion 1; Assertion 2 is immediate from the definition. We note that this result identifies $H^1(M, \mathbb{Z}_2)$ with $\mathrm{Hom}(\pi_1(M); \mathbb{Z}_2)$. □

18.3.4 THE SECOND STIEFEL–WHITNEY CLASS. Let V be an oriented vector bundle of dimension ℓ over a compact manifold M. The first Stiefel–Whitney class is the obstruction to orienting V. The second Stiefel–Whitney class is the obstruction to finding a spin structure on V. It is defined as follows. Fix a fiber metric h on V. Choose local oriented orthonormal frames $\vec{s}_i$ for V; the transition functions $\phi(i,j)$ then belong to $\mathrm{SO}(\ell)$. Let $\varrho : \mathrm{Spin}(\ell) \to \mathrm{SO}(\ell)$ be as discussed in Section 18.3.2. We say that (V, h) admits a *spin structure* if it is possible to find $\xi(i,j) : \mathcal{O}_i \cap \mathcal{O}_j \to \mathrm{Spin}(\ell)$ so that $\varrho(\xi(i,j)) = \phi(i,j)$ and so that $\xi(i,j)\xi(j,k)\xi(k,i) = 1$ so ξ satisfies the cocycle condition. We will show in Example 18.22 that there can be inequivalent spin structures. The collection of functions $\{\phi(i,j)\}$ satisfies the cocycle condition.

Define an element of $\mathcal{C}^2(M, \mathbb{Z}_2)$ by setting $\varepsilon(i,j,k) := \xi(i,j)\xi(j,k)\xi(k,i)$. We then have $\varrho\{\varepsilon(i,j,k)\} = \mathrm{Id}$. Consequently, Lemma 18.15 shows that $\varepsilon(i,j,k) = \pm 1$ and ε belongs to $\mathcal{C}^2(M, \mathbb{Z}_2)$. It is a lengthy and not terribly enlightening computation to show $\delta(\varepsilon)$ is trivial so ε determines an element $[\varepsilon] \in H^2(M, \mathbb{Z}_2)$. Furthermore, $[\varepsilon]$ is independent of the choices made and defines an element we shall denote by $\mathrm{sw}_2(V) \in H^2(M, \mathbb{Z}_2)$. This shows:

Lemma 18.19   Let (V, h) be an oriented real vector bundle over a compact manifold M. Then (V, h) admits a spin structure if and only if $\mathrm{sw}_2(V) = 0$ in $H^2(M, \mathbb{Z}_2)$.

We say that a Riemannian manifold (M, g) is *orientable* if TM is orientable. If M is orientable, we say that M is a *spin manifold* if TM admits a spin structure. Let $\mathbb{RP}^m$ be the set of real lines through the origin in $\mathbb{R}^{m+1}$; we can identify $\mathbb{RP}^m$ with $S^m/\{\pm 1\}$ where we identify antipodal points of the sphere. The real *tautological line bundle* $\mathbb{L}_\mathbb{R}$ over $\mathbb{RP}^m$ associates to each "point" of $\mathbb{RP}^m$ the real line which it represents. More precisely, $\mathbb{L}_\mathbb{R} := S^m \times \mathbb{R}/\sim$ where we identify (θ, λ) with $(-\theta, -\lambda)$. Similarly, let $\mathbb{CP}^m$ be the set of complex lines through the origin in $\mathbb{C}^{m+1}$; we refer to Section 19.2.1 for further details. We can identify $\mathbb{CP}^m$ with S^{2m+1}/S^1 in the Hopf fibration. Let $\mathbb{L}_\mathbb{C}$ be the tautological complex line bundle over $\mathbb{CP}^m$ that associates to each "point" of $\mathbb{CP}^m$ the corresponding complex line which it represents; we refer to

Section 19.2.2 for details. If V is a vector bundle over N and if $f : M \to N$, let $f^*(V)$ be the pullback bundle over M; we refer to Section 5.4.1 of Book II for details.

Example 18.20

1. The sphere S^m. Let ν be the normal bundle of S^m in $\mathbb{R}^{m+1}$. Since ν is trivial, the decomposition $T(S^m) \oplus \nu = T(\mathbb{R}^{m+1})|_{S^m}$ shows that $T(S^m)$ is stably trivial, i.e.,

$$T(S^m) \oplus \mathbf{1} = (m+1) \cdot \mathbf{1}.$$

   Thus, $\mathrm{sw}_k(S^m) = 0$ for $k > 0$; S^m is orientable and spin.

2. Real projective space $\mathbb{RP}^m$. $H^*(\mathbb{RP}^m; \mathbb{F}_2) = \mathbb{F}_2[x_1]/(x_1^{m+1} = 0)$ is a truncated polynomial ring where $x_1 := \mathrm{sw}_1(\mathbb{L}_\mathbb{R})$ generates $H^1(\mathbb{RP}^m; \mathbb{F}_2) = \mathbb{F}_2$. The decomposition

$$T(S^m) \oplus \mathbf{1} = T(\mathbb{R}^{m+1})|_{S^m}$$

   is $\mathbb{F}_2$ equivariant and defines a corresponding isomorphism

$$T\mathbb{RP}^m \oplus \mathbf{1} = (m+1)\mathbb{L}_\mathbb{R}.$$

   Consequently, $\mathrm{sw}(T\mathbb{RP}^m) = \mathrm{sw}(T\mathbb{RP}^m \oplus \mathbf{1}) = (1+x_1)^{m+1}$.

   (a) $\mathrm{sw}_1(T\mathbb{RP}^m) = (m+1)x_1$; $\mathbb{RP}^m$ is orientable if and only if m is odd.

   (b) $\mathrm{sw}_2(T\mathbb{RP}^m) = \frac{1}{2}(m+1)m\, x_1^2$; $\mathbb{RP}^m$ is spin if and only if $m \equiv 3 \bmod 4$.

3. Complex projective space $\mathbb{CP}^m$. $H^*(\mathbb{CP}^m; \mathbb{F}_2) = \mathbb{F}_2[x_2]/(x_2^{m+1} = 1)$ is a truncated polynomial ring where $x_2 := \mathrm{sw}_2(\mathbb{L}_\mathbb{C})$ generates $H^2(\mathbb{CP}^m; \mathbb{F}_2) = \mathbb{F}_2$. We will show in Lemma 19.9 that $T_c(\mathbb{CP}^m) \oplus \mathbf{1} = (m+1)\mathbb{L}_\mathbb{C}$ so $\mathrm{sw}(TM) = (1+x_2)^{m+1}$.

   (a) $\mathrm{sw}_1(\mathbb{CP}^m)$ is trivial so $\mathbb{CP}^m$ is orientable for arbitrary m.

   (b) $\mathrm{sw}_2(\mathbb{CP}^m) = (m+1)x_2$ so $\mathbb{CP}^m$ is spin if and only if m is odd.

18.3.5 THE LICHNEROWICZ FORMULA [47]. Let $\mathrm{Cliff}(M)$ be the Clifford algebra of the tangent bundle of M; we suppress the dependence upon the Riemannian metric g as it does not affect the isomorphism class of the bundle. Let $\tau = R_{ijji}$ be the *scalar curvature*, let Γ_{ijk} be the *Christoffel symbols*, and let $R_{ij} : e_k \to R_{ijk\ell}e_\ell$ be the *curvature operator* of the Levi–Civita connection expressed with respect to a local orthonormal frame of the tangent bundle. The following result is the appropriate generalization of the Weitzenböck formula (see Lemma 18.8) to spinor setting.

Lemma 18.21 Let (M, g) be a compact Riemannian manifold of dimension $m = 2\mathfrak{m}$ or of dimension $m = 2\mathfrak{m} + 1$ which is equipped with a spin structure.

1. The spin bundle $\mathcal{S}(M)$ is a complex vector bundle over M of fiber dimension $2^{\mathfrak{m}}$ which has a Hermitian metric $h_\mathcal{S}$, a Clifford module structure $c_\mathcal{S}$, and a connection $\nabla_\mathcal{S}$ so that:

   (a) $c_S(\xi)$ is skew-adjoint for every $\xi \in T^*M$.

   (b) $\nabla_S c_S = 0$. The connection one-form of ∇_S is $\frac{1}{4}\Gamma_{ijk}c_S(e_j)c_S(e_k)$.

   (c) The curvature operator $R_S(e_i, e_j)$ of ∇_S is $\frac{1}{4}R_{ijk\ell}c_S(e_k * e_\ell)$.

2. The spin-Dirac operator $A_S := C_S \circ \nabla_S$ is a self-adjoint operator of Dirac type.

3. The spin Laplacian $\Delta_S := A_S^2$ is a self-adjoint operator of Laplace type which satisfies the
   identity $\Delta_S \theta = -\theta_{;ii} + \frac{1}{4}\tau\theta$.

Proof. Let $\{\mathcal{O}_1, \dots, \mathcal{O}_n\}$ cover M by small geodesic balls, let $\phi_{ij} : \mathcal{O}_i \cap \mathcal{O}_j \to SO(m)$ be the
transition functions of the tangent bundle, and let $\xi_{ij} : \mathcal{O}_i \cap \mathcal{O}_j \to \text{Spin}(m)$ define the spin
structure; $\varrho(\xi_{ij}) = \phi_{ij}$ and $\xi_{ij} * \xi_{jk} * \xi_{ki} = \text{Id}$. We adopt the notation of Lemma 18.4. Let
$S = S_m$ if m is even and $S = S_m^+$ if m is odd; S is a $\text{Cliff}(\mathbb{R}^m)$ module and, consequently, S
is a representation space for $\text{Spin}(m)$. Since $\xi_{ij} * \xi_{jk} * \xi_{ki} = 1$, $c_S(\xi_{ij})c_S(\xi_{jk})c_S(\xi_{ki}) = \text{Id}$, the
collection $\{c_S(\xi_{ij})\}$ satisfies the cocycle condition and defines the *spin bundle* $S(M)$ over M.
The representation S comes equipped with a Hermitian inner product h_S; $c_S(\xi_{ij})$ preserves this
inner product. Consequently, h_S extends to a Hermitian inner product on $S(M)$. We have a
commutative diagram

$$
\begin{array}{ccc}
\xi \otimes \theta & \to & c_S(\xi) * \theta \\
\downarrow & \circ & \downarrow \\
(\xi_{ij} * \xi * \xi_{ij}^{-1}) \otimes (c_S(\xi_{ij})\theta) & \to & c_S(\xi_{ij})c_S(\xi)c_S(\xi_{ij}^{-1})c_S(\xi_{ij})\theta = c_S(\xi_{ij})c_S(\xi)\theta\,.
\end{array}
$$

Consequently, Clifford multiplication c_S is compatible with the gluing maps and we have a well
defined map $c_S : \text{Cliff}(TM, g) \to \text{End}(S(M))$. The connection one-form ω_i for the Levi–Civita
connection on the tangent bundle is $\frac{1}{2}\Gamma_{ijk}\mathfrak{so}_{jk} \otimes e_i$ where $\mathfrak{so}_{jk}$ is defined by Equation (18.3.d).
We use Lemma 18.16 to lift this to the Lie algebra of Spin and define $\omega_{i,\text{Spin}} := \frac{1}{4}\Gamma_{ijk}c(e_j)c(e_k)$
to obtain a Hermitian connection ∇_S on $S(M)$. Equation (18.1.e) then shows $\nabla_S\{c_S\} = 0$. This
establishes Assertion 1.

   We use Lemma 18.7 to define the *spin Dirac operator* $A_S := c_S \circ \nabla_S$. Assertion 2 follows
from Assertion 1 and from Lemma 18.7. We may also conclude

$$
\Delta_S\theta = -\theta_{;ii} + \frac{1}{8}R_{ijk\ell}c_S(e_i * e_j * e_k * e_\ell)R_{ijk\ell}
$$

is self-adjoint. As in the proof of Lemma 18.8, we may use the first Bianchi identity to see
$\frac{1}{8}R_{ijk\ell}c_S(e_i * e_j * e_k) = 0$ if i, j, and k are distinct indices. We therefore set $j = k$ to complete
the proof of Assertion 3 by computing:

$$
\frac{1}{8}R_{ijk\ell}c_S(e_i * e_j * e_k * e_\ell) = -\frac{1}{4}R_{ijj\ell}c_S(e_i * e_\ell) = \frac{1}{4}\tau\,\text{Id}\,. \qquad \square
$$

   There can be inequivalent spin structures. Let $[\varepsilon] \in H^1(M, \mathbb{Z}_2)$. Then $\varepsilon(i, j)$ is a collection
of $\pm$ signs so that $\varepsilon(i, j)\varepsilon(j, k)\varepsilon(k, i) = 1$. If $\{\xi(i, j)\}$ is a lift of the transition functions of V
from $SO(V, h)$ to $\text{Spin}(V, h)$ satisfying the cocycle condition, then $\{\varepsilon(i, j)\xi_S(i, j)\}$ also is a lift

of the transition functions to $\mathrm{Spin}(V, h)$ which satisfies the cocycle condition; this defines the same spin structure if and only if $[\varepsilon] = 0$ in $H^2(M, \mathbb{Z})$.

Example 18.22

1. Let $M = [0, 2\pi]/ \sim$ where we identify 0 with 2π to obtain the circle. The tangent bundle is the trivial bundle. Let $\mathcal{O}_1 := (-\frac{\pi}{2}, \frac{\pi}{2})$, $\mathcal{O}_2 := (0, \pi)$, $\mathcal{O}_3 := (\frac{\pi}{2}, \frac{3\pi}{2})$, and $\mathcal{O}_4 := (\pi, 2\pi)$ be a simple cover of M. We regard the index i as defined modulo 4; $\mathcal{O}_i \cap \mathcal{O}_j$ is non-empty if and only if $i \neq j + 2$. We take the transition functions $\phi_{ij} = 1$. Since

$$H^1(S^1; \mathbb{Z}_2) = \mathrm{Hom}(\pi_1(S^1), \mathbb{Z}_2) = \mathrm{Hom}(\mathbb{Z}, \mathbb{Z}_2) = \mathbb{Z}_2,$$

   there are two inequivalent spin structures on S^1.

   (a) Let $\xi_{ij} = 1$ for all i define the spin structure $\mathcal{S}_1$. The associated spin bundle is $S^1 \times \mathbb{R}$ and the associated Laplacian is $-\partial_x^2$. Consequently, $\{e^{\sqrt{-1}n\theta}, n^2\}_{n \in \mathbb{Z}}$ is a complete spectral resolution of the spin Laplacian and $\dim(\ker\{\Delta_{\mathcal{S}_1}\}) = 1$.

   (b) Let $\xi_{34} = -1$ and $\xi_{ij} = +1$ otherwise. This satisfies the cocycle condition since we have that $\mathcal{O}_i \cap \mathcal{O}_j \cap \mathcal{O}_k$ is empty if $\{i, j, k\}$ are distinct and we obtain a spin structure $\mathcal{S}_2$. The associated real line bundle $\Delta_{\mathcal{S}_2}$ is the Möbius strip over the circle; $\xi_{34} = -1$ puts a single twist in the bundle. Thus, these two structures are distinct. Again the structures are flat and the Laplacian is $-\partial_x^2$. However, $\{e^{\sqrt{-1}(n+\frac{1}{2})\theta}, (n + \frac{1}{2})^2\}_{n \in \mathbb{Z}}$ is a complete spectral resolution of $\Delta_{\mathcal{S}_2}$; we need to ensure that $\phi(0) = -\phi(2\pi)$ to ensure that ϕ is a section to the Möbius bundle. In particular, $\dim(\ker\{\Delta_{\mathcal{S}_2}\}) = 0$.

2. More generally, if $\mathbb{T}^k = S^1 \times \cdots \times S^k$ with the flat metric, there will be 2^k distinct spin structures giving rise to 2^k spin bundles. Exactly one of these structures gives rise to a Laplace operator where $\dim(\ker\{\Delta_{\mathcal{S}}\}) = 1$; for the remaining spin structures $\dim(\ker\{\Delta_{\mathcal{S}}\}) = 0$. For exactly one of these structures, the smallest eigenvalue of the spin Laplacian is $\frac{k}{4}$; for the remaining $2^k - 2$ spin structures the smallest eigenvalue of the spin Laplacian will range between $\frac{1}{4}$ and $\frac{k-1}{4}$.

18.3.6 OBSTRUCTIONS TO POSITIVE SCALAR CURVATURE.

Suppose that M has dimension $m = 2\mathrm{m}$ and admits a spin structure. The orientation defines an element orn of $C^\infty(\Lambda^m M) = C^\infty(\mathrm{Cliff}(M))$. Since $c_{\mathcal{S}}((\sqrt{-1})^\mathrm{m}\mathrm{orn})^2 = \mathrm{Id}$, we may decompose the spinor bundle into the $\pm$ eigenspaces $\mathcal{S}_\pm(M)$ of $c_{\mathcal{S}}((\sqrt{-1})^\mathrm{m}\mathrm{orn})$. If v is a tangent vector, then $c_{\mathcal{S}}(v)$ anti-commutes with $c_{\mathcal{S}}((\sqrt{-1})^\mathrm{m}\mathrm{orn})$. Consequently, the spin Dirac operator $A_{\mathcal{S}}$ decomposes as $A_{\mathcal{S},+} \oplus A_{\mathcal{S},-}$ where

$$A_{\mathcal{S},\pm} : C^\infty(\mathcal{S}_\pm(M)) \to C^\infty(\mathcal{S}_\mp(M)). \tag{18.3.e}$$

Equation (18.3.e) defines the *spin complex* $\mathcal{E}_{\mathrm{spin}}$. This is an elliptic complex of Dirac type. Let $\hat{A}$ be the A-hat genus. It is defined in terms of the Pontrjagin classes in Section 16.1.6.f. The following result is due to Lichnerowicz [47] and generalizes the Bochner Vanishing Theorem of Theorem 18.11 to the context of spin geometry.

Theorem 18.23   Let (M, g) be a compact connected Riemannian spin manifold of dimension m. Suppose that $\tau \geq 0$ and that τ is positive at some point of M.

1. $\dim(\ker\{\Delta_S\}) = 0$.
2. If $m = 4m$ is divisible by 4, then $\text{Index}(\mathcal{E}_{\text{spin}}) = \int_M \hat{A}_{4k}(TM) = 0$.

Proof.   Let $\theta \in \ker\{\Delta_S\}$. We use Lemmas 16.4 and 18.21 to compute

$$0 = (\Delta_S\theta, \theta)_{L^2} = (\nabla_S\theta, \nabla_S\theta)_{L^2} + (\tfrac{1}{4}\tau\theta, \theta)_{L^2}.$$

This implies $\|\nabla_S\theta\|_{L^2}^2 = 0$ so θ is parallel. Consequently, $\|\theta\|^2$ is constant. Since τ is non-negative and τ is positive at some point, the vanishing of $(\tfrac{1}{4}\tau\theta, \theta)_{L^2}$ implies $\theta = 0$. This proves Assertion 1; Assertion 2 follows from Assertion 1 and from the Atiyah–Singer Index Theorem.   □

Remark 18.24   The $\hat{A}$-genus is defined by the underlying topology of the manifold. If $m \equiv 0$ mod 4 and if $\int_M \hat{A}_m \neq 0$, Theorem 18.23 shows that M does not admit a metric of positive scalar curvature; the $\hat{A}$ genus is an obstruction to the existence of a metric of positive scalar curvature in the spin context in dimensions divisible by 4. Consider, for example, the Fermat variety given projectively in $\mathbb{CP}^3$ by the homogeneous equations

$$M := \{\langle z^1, z^2, z^3, z^4\rangle \in \mathbb{C}^4 - \{0\} : z_1^4 + z_2^4 + z_3^4 + z_4^4 = 0\}/\{\mathbb{C} - \{0\}\}.$$

This *K3 surface* was first examined by Kummer. It is spin and $\hat{A}_4$ is non-zero. Consequently, it does not admit a metric of positive scalar curvature. We shall discuss K3 surfaces in a bit more detail in Remark 19.15 subsequently. Theorem 18.23 is valid only in the spin context. We have that $\hat{A}_4[\mathbb{CP}^2]$ is non-zero and the Fubini–Study metric on $\mathbb{CP}^2$ has positive scalar curvature. However, Theorem 18.23 does not apply since $\mathbb{CP}^2$ is not spin (see Example 18.20).

By now, there are many applications of the Lichnerowicz principle to construct obstructions to finding metrics of positive scalar curvature; we refer to the historical survey in Gilkey [28], for example, as limitations of space do not permit a further discussion here. We content ourselves with presenting an argument due to Gromov and Lawson [32] that shows that the torus $\mathbb{T}^m := \mathbb{R}^m/\mathbb{Z}^m$ does not admit a metric of positive scalar curvature.

Let (E, h, ∇) be a Hermitian vector bundle which is equipped with a unitary connection ∇. Let $\nabla_{S\otimes E}$ be the associated connection on $\mathcal{S} \otimes E$. We define the spin complex with coefficients in E by $A_{S,E} := c \circ \nabla_{S\otimes E} : C^\infty(\mathcal{S}_\pm \otimes E) \to C^\infty(\mathcal{S}_\mp \otimes E)$ with associated Laplacian $\Delta_{S,E} = -g^{ij}\nabla_i\nabla_j + \tfrac{1}{4}\tau \otimes \text{Id}_E + \text{Id}_S \otimes R_E$ where R_E is the curvature operator of ∇_E. We then obtain:

Theorem 18.25

1. Let (M, g) be a compact connected Riemannian spin manifold with $\tau \geq \varepsilon > 0$. There exists $\delta = \delta(\varepsilon)$ so that if $\|R_E\|^2 \leq \delta$, then $\mathrm{Index}(\mathcal{E}_{\mathrm{Spin},E}) = 0$.

2. The torus $\mathbb{T}^m$ does not admit a metric of positive scalar curvature.

Proof. The first assertion is immediate. If $\mathbb{T}^m$ admits a metric of positive scalar curvature τ, then by taking the Cartesian product with the circle, we conclude $\mathbb{T}^{m+1}$ admits a metric of positive scalar curvature. Thus, we may increase the dimension if necessary to assume $m = 2\mathrm{m}$ is even. We apply Assertion 1. Let $\mathbb{T}_k^m := \mathbb{R}^m / k \cdot \mathbb{Z}^m$ for $m = 2\mathrm{m}$. Suppose that $\mathbb{T}_1^m$ for $m = 2\mathrm{m}$ admits a metric of positive scalar curvature with $\tau \geq \varepsilon > 0$. Choose $\kappa > 0$ so that $|g_{ij}| \leq \kappa$ and $|g^{ij}| \leq \kappa$. The metric on $\mathbb{T}_1^m$ defines a $\mathbb{Z}^m$-invariant metric on $\mathbb{R}^m$ which then restricts to a metric on $\mathbb{T}_k^m$ satisfying $\tau \geq \varepsilon > 0$ and so $g_{ij} \leq \kappa$ and $g^{ij} \leq \kappa$. Let ι be the injectivity radius; $\iota(\mathbb{T}_k^m) = k\iota(\mathbb{T}_1^m)$. Thus, the injectivity radius can be made arbitrarily large.

By Theorem 16.12, we can find a geometrical vector bundle (E, h, ∇) over S^m with $\int_{S^m} \mathrm{ch}_{\mathrm{m}}(E) \neq 0$. Let $F : \mathbb{T}_k^m \to S^m$ be a degree one mapping. If k is sufficiently large, then $\|F^* R_E\|^2 < \delta$. Consequently, $\mathrm{Index}(\mathcal{E}_{\mathrm{Spin}, F^*E}) = 0$. All the Pontrjagin classes of the torus vanish since it admits a flat metric. Consequently, the Atiyah–Singer Index Theorem yields contradiction completes the proof:

$$\mathrm{Index}(\mathcal{E}_{\mathrm{Spin}, F^*E}) = \int_{\mathbb{T}_k^m} \mathrm{ch}_{\mathrm{m}}(F^*E) = \int_{S^m} \mathrm{ch}_{\mathrm{m}}(E) \neq 0. \qquad \square$$

Remark 18.26 The torus is *enlargeable* in the sense of Gromov and Lawson [32] and the argument given above shows any enlargeable manifold does not admit a metric of positive scalar curvature. If $m = 2$, it is possible to give a more direct argument using the Gauss–Bonnet Theorem to derive the desired contradiction:

$$0 = \chi(\mathbb{T}^2) = \frac{1}{4\pi} \int_{\mathbb{T}^2} \tau \geq \frac{\varepsilon}{4\pi} \mathrm{Vol}(\mathbb{T}^2) > 0.$$

18.4   THE DOLBEAULT COMPLEX

In Section 18.4.1, we treat the Cauchy–Riemannian equations. In Sections 18.4.2 and 18.4.3, we derive the holomorphic inverse and implicit function theorems from the corresponding real analogues. In Section 18.4.4, we define the notion of a holomorphic structure and present the Newlander–Nirenberg Integrability Theorem. We introduce the bundles $\Lambda^{p,q}$ in Section 18.4.5. We show the Dolbeault complex is an elliptic complex of Dirac type in Section 18.4.6. We discuss the Chern connection in Section 18.4.7, the Riemann–Roch Theorem in Section 18.4.8, and the Hirzebruch–Riemann–Roch Theorem in Section 18.4.9.

18.4.1 THE CAUCHY–RIEMANN EQUATIONS. We begin by reviewing a foundational result from the theory of complex variables. Let $\mathcal{O}$ be an open subset of $\mathbb{C}$. A function

$$f : \mathcal{O} \to \mathbb{C}$$

is said to be *holomorphic* on $\mathcal{O}$ if $f \in C^2(\mathcal{O})$ and if the complex derivative

$$f'(z) = \partial_z f := \lim_{\delta \to 0} \delta^{-1}\{f(z + \delta) - f(z)\}$$

exists for every point $z \in \mathcal{O}$; we emphasize that δ is complex in taking this limit. Let $\mathrm{Hol}(\mathcal{O})$ be the set of holomorphic functions which are defined on $\mathcal{O}$. The arguments in real single-variable calculus extend to show $\mathrm{Hol}(\mathcal{O})$ is a unital ring under pointwise sums and products; furthermore, if $g \in \mathrm{Hol}(\mathcal{O})$ never vanishes, then $g^{-1} \in \mathrm{Hol}(\mathcal{O})$. The following is well known; we give the proof for the sake of completeness. We remark that the equations $u_x = v_y$ and $u_y = -v_x$ are called the *Cauchy–Riemann Equations*.

Lemma 18.27   Let $\mathcal{O}$ be an open subset of $\mathbb{C}$. Let $f = u + \sqrt{-1}v$ define a map from $\mathcal{O}$ to $\mathbb{C}$ where u and v are real-valued C^2 functions.

1. $f \in \mathrm{Hol}(\mathcal{O})$ if and only if $u_x = v_y$ and $u_y = -v_x$. We have $f'(z) = u_x + \sqrt{-1}v_x$.
2. Let $f_r = (u, v) : \mathcal{O} \to \mathbb{R}^2$ be the underlying real map. Then $\det(f_r') = |f'|^2$.
3. If $f \in \mathrm{Hol}(\mathcal{O})$ then f is C^∞ and $\partial_z^k f \in \mathrm{Hol}(\mathcal{O})$ for any k.
4. $f \in \mathrm{Hol}(\mathcal{O})$ if and only if $d(f(z)dz) = 0$.

Proof. Let $f \in \mathrm{Hol}(\mathcal{O})$. We compute f' by taking the limit along the real axis ($\delta = t$) and the purely imaginary axis ($\delta = \sqrt{-1}t$) to see

$$
\begin{aligned}
f'(z) &= \lim_{t \to 0} t^{-1}\{f(x + t, y) - f(x, y)\} = u_x + \sqrt{-1}v_x \\
&= \lim_{t \to 0} (\sqrt{-1}t)^{-1}\{f(x, y + t) - f(x, y)\} \\
&= -\sqrt{-1}\{u_y + \sqrt{-1}v_y\} = v_y - \sqrt{-1}u_y .
\end{aligned}
$$

Equating real and imaginary parts yields the Cauchy–Riemann equations. We also may conclude that $f' = u_x + \sqrt{-1}v_x$. Conversely, assume $u_x = v_y$ and $u_y = -v_x$. Let $f_r = (u, v)$;

$$f_r' = \begin{pmatrix} u_x & u_y \\ v_x & v_y \end{pmatrix} = \begin{pmatrix} u_x & -v_x \\ v_x & u_x \end{pmatrix}.$$

Let $\delta = \delta_x + \sqrt{-1}\delta_y$. Since f_r is C^1, we complete the proof of Assertion 1 by verifying:

$$
\begin{aligned}
0 &= \lim_{\delta \to 0} \|\delta\|^{-1} \|f(z + \delta) - f(z) - f_r'(z)(\delta)\| \\
&= \lim_{\delta \to 0} \|\delta\|^{-1} \|f(z + \delta) - f(z) - (u_x\delta_x - v_x\delta_y) - \sqrt{-1}(v_x\delta_x + u_x\delta_y)\|
\end{aligned}
$$

$$= \lim_{\delta \to 0} \|\delta\|^{-1} \|f(z+\delta) - f(z) - (u_x + \sqrt{-1}v_x)(\delta_x + \sqrt{-1}\delta_y)\|$$

$$= \lim_{\delta \to 0} \|\delta\|^{-1} \|f(z+\delta) - f(z) - (u_x + \sqrt{-1}v_x)\delta\|$$

$$= \lim_{\delta \to 0} \|\delta^{-1}\{f(z+\delta) - f(z)\} - (u_x + \sqrt{-1}v_x)\| \,.$$

We use Assertion 1 to prove Assertion 2 by computing:

$$\det(f_r') = \det \begin{pmatrix} u_x & u_y \\ v_x & v_y \end{pmatrix} = \det \begin{pmatrix} u_x & u_y \\ -u_y & u_x \end{pmatrix} = u_x^2 + u_y^2 = |f'|^2 \,.$$

Suppose $f \in \mathrm{Hol}(\mathcal{O})$. Then $u_x = v_y$ and $u_y = -v_x$. We compute therefore

$$u_{xx} + u_{yy} = u_{xx} + v_{xy} = u_{xx} - u_{xx} = 0,$$

$$v_{xx} + v_{yy} = v_{xx} + u_{xy} = v_{xx} - v_{xx} = 0 \,.$$

Consequently, both u and v are harmonic. Fix a point $P \in \mathcal{O}$ and let $\psi \in C_0^\infty(\mathcal{O})$ be a plateau function which is identically 1 on a smaller neighborhood $\mathcal{U}$ of P. Embed $\mathcal{O}$ holomorphically as an open subset of the torus. Let $\Delta := -\partial_x^2 - \partial_y^2$ be the Laplacian on the torus. We can assume ψ has support in the torus. Then $\Delta\psi u = 0$ on $\mathcal{U}$ so Theorem 17.23 shows ψu is smooth on $\mathcal{U}$ and hence u is smooth on $\mathcal{U}$. Since P was arbitrary, $u \in C^\infty(\mathcal{O})$. The argument is the same for v. Since the Cauchy–Riemann equations are constant coefficient partial differential equations, they hold for all the derivatives of f and hence $\partial_z^k f \in \mathrm{Hol}(\mathcal{O})$. We remark that we will give another proof of this smoothness result in Chapter 19 when we discuss Taylor series in Lemma 19.2. Assertion 4 follows from the computation:

$$d(f(z)dz) = d\{(u + \sqrt{-1}v)(dx + \sqrt{-1}dy)\}$$

$$= \{(-v_x - u_y) + \sqrt{-1}(u_x - v_y)\}dx \wedge dy \,. \qquad \square$$

18.4.2 THE HOLOMORPHIC INVERSE FUNCTION THEOREM. The (real) Inverse Function Theorem and Implicit Function Theorem (see Theorems 1.8 and 1.10 of Book I) play a central role in real Differential Geometry. There are corresponding holomorphic analogues which are useful in the complex setting. If $\mathcal{O}$ is an open subset of $\mathbb{C}^m$, then we say that $\vec{F} = (f^1, \dots, f^m) : \mathcal{O} \to \mathbb{C}^m$ is *holomorphic* if the component functions f^1, ..., f^m are holomorphic. Let $\vec{F}_c' := \left(\frac{\partial f^i}{\partial z^j}\right)$ be the complex Jacobian.

Theorem 18.28 Let $\mathcal{O}$ be an open subset of $\mathbb{C}^m$. Let $\vec{F} = (f^1, \dots, f^m) : \mathcal{O} \to \mathbb{C}^m$ be holomorphic. Assume $\det_{\mathbb{C}}(\vec{F}_c'(\vec{z}_0)) \neq 0$ for some $\vec{z}_0 \in \mathcal{O}$. There exists a smaller neighborhood $\tilde{\mathcal{O}}$ of $\vec{z}_0$ contained in $\mathcal{O}$ so that $\mathcal{U} := \vec{F}(\tilde{\mathcal{O}})$ is an open neighborhood of $F(\vec{z}_0)$, so that $\vec{F} : \tilde{\mathcal{O}} \to \mathcal{U}$ is 1-1 and onto, and so that $\vec{F}^{-1} : \mathcal{U} \to \tilde{\mathcal{O}}$ is 1-1 and holomorphic.

Proof. If A_c is an $\mathfrak{m} \times \mathfrak{m}$ complex matrix, let A_r be the corresponding underlying real $2\mathfrak{m} \times 2\mathfrak{m}$ matrix. If A_c is diagonalizable, let $\{\lambda_i = a_i + \sqrt{-1}b_i\}$ be the complex eigenvalues of A_c. We then have $\det(A_c) = \lambda_1 \cdot \cdots \cdot \lambda_\mathfrak{m}$. Since $\det(\bar{A}_c) = \bar{\lambda}_1 \cdot \cdots \cdot \bar{\lambda}_\mathfrak{m}$, we have

$$\det(A_c)\det(\bar{A}_c) = \lambda_1\bar{\lambda}_1 \cdot \cdots \cdot \lambda_\mathfrak{m}\lambda_\mathfrak{m} = (a_1^2 + b_1^2) \cdot \cdots \cdot (a_\mathfrak{m}^2 + b_\mathfrak{m}^2).$$

The real form A_r decomposes into 2×2 blocks

$$A_r = A_1 \oplus \cdots \oplus A_\mathfrak{m} \quad \text{where} \quad A_i = \begin{pmatrix} a_i & b_i \\ -b_i & a_i \end{pmatrix}.$$

Since $\det(A_i) = a_i^2 + b_i^2$, $\det(A_c)\det(\bar{A}_c) = \det(A_r)$ if A_c is complex diagonal. Since the determinant is independent of the particular basis chosen, $\det(A_c)\det(\bar{A}_c) = \det(A_r)$ if A_c is complex diagonalizable. As the complex diagonalizable matrices are dense in the space of $\mathfrak{m} \times \mathfrak{m}$ complex matrices, $|\det(A_c)|^2 = \det(A_r)$.

Let $\vec{F}_r : \mathcal{O} \to \mathbb{R}^{2\mathfrak{m}}$ be the underlying real map which corresponds to a holomorphic map $\vec{F}_c : \mathcal{O} \to \mathbb{C}^\mathfrak{m}$. In one variable, Lemma 18.27 shows $\det(\vec{F}_r') = |\det(\vec{F}_c')|^2$. We show that $\det(\vec{F}_r') = |\det(\vec{F}_c')|^2$ in general by arguing as above. Consequently, $\vec{F}_r$ satisfies hypotheses of the Inverse Function Theorem (Theorem 1.8 of Book I). Thus, $\vec{F}_r^{-1}$ exists on some smaller subdomain and is C^2. We have $(\vec{F}_r^{-1})' = (\vec{F}_r')^{-1} \circ \vec{F}_r^{-1}$. It now follows that $(\vec{F}_r^{-1})'$ is the underlying real form of the matrix $(F_c')^{-1}$ composed with $\vec{F}_r^{-1}$ and hence satisfies the Cauchy–Riemann equations. This proves $\vec{F}^{-1}$ is holomorphic. □

18.4.3 THE HOLOMORPHIC IMPLICIT FUNCTION THEOREM.

Theorem 18.29

1. Let $\vec{z} = (z^1, \dots, z^a) \in \mathcal{O}_1$ and $\vec{w} = (w^1, \dots, w^b) \in \mathcal{O}_2$ where $\mathcal{O}_1$ is an open subset of $\mathbb{C}^a$ and $\mathcal{O}_2$ is an open subset of $\mathbb{C}^b$. Let $\vec{F} = (f^1, \dots, f^a)(\vec{z}, \vec{w})$ be holomorphic on the open set $\mathcal{O}_1 \times \mathcal{O}_2$. Assume $\det_{\mathbb{C}}(\frac{\partial f^i}{\partial z^j})(\vec{z}_0, \vec{w}_0) \neq 0$ for some point $(\vec{z}_0, \vec{w}_0) \in \mathcal{O}_1 \times \mathcal{O}_2$. There exists a smaller neighborhood $\tilde{\mathcal{O}}_1$ of $\vec{z}_0$, which is contained in $\mathcal{O}_1$, and a smaller neighborhood $\tilde{\mathcal{O}}_2$ of $\vec{w}_0$ contained in $\mathcal{O}_2$ and holomorphic functions $z^i(w^1, \dots, w^b)$ for $1 \leq i \leq a$ so that $z^i(\vec{w}_0) = z_0^i$ and so that $\vec{F}(\vec{z}, \vec{w}) = \vec{F}(\vec{z}_0, \vec{w}_0)$ if and only if $\vec{z} = \vec{z}(\vec{w})$.

2. Let $\mathcal{U}$ be an open subset of $\mathbb{C}^\mathfrak{m}$. Let $\vec{F} = (f^1, \dots, f^\mathfrak{n})$ be a holomorphic map from $\mathcal{U}$ to $\mathbb{C}^\mathfrak{n}$. Let P be a point of U with $\{\partial f^1 \wedge \cdots \wedge \partial f^\mathfrak{n}\}(P) \neq 0$. Then there exists a smaller neighborhood $\tilde{\mathcal{U}}$ of P contained in $\mathcal{U}$ so that $\{\vec{z} \in \tilde{\mathcal{U}} : \vec{F}(\vec{z}) = \vec{F}(P)\}$ is a holomorphic submanifold of $\tilde{\mathcal{U}}$ of complex dimension $\mathfrak{m} - \mathfrak{n}$.

18.4.4 THE NEWLANDER–NIRENBERG THEOREM [52].
We refer to Section 4.3 of Book II for further details concerning the material of this section. Let M be a manifold of

real dimension 2m. An endomorphism J of TM with $J^2 = -\operatorname{Id}$ is said to be an *almost complex structure*. A coordinate system $(x^1, \ldots, x^m, y^1, \ldots, y^m)$ is said to be a *holomorphic coordinate system* if $J(\partial_{x^i}) = \partial_{y^i}$ and $J(\partial_{y^i}) = -\partial_{x^i}$. Set $z^i := x^i + \sqrt{-1} y^i$. Let

$$N_J(X, Y) := [X, Y] + J([JX, Y]) + J([X, JY]) - [JX, JY]$$

be the *Nijenhuis tensor*. A holomorphic structure on M is an atlas $\{\mathcal{O}_\alpha, \phi_\alpha\}$ for M with holomorphic transition functions $\phi_\alpha \circ \phi_\beta^{-1}$ where $\phi_\alpha : \mathcal{O}_\alpha \to \mathcal{U}_\alpha \subset \mathbb{C}^m$.

Theorem 18.30

1. Every holomorphic manifold is equipped with a natural almost complex structure J so that if $(z^1, \ldots, z^m)$ is a system of local holomorphic coordinates on M, then $J\partial_{x^i} = \partial_{y^i}$ and $J\partial_{y^i} = -\partial_{x^i}$.

2. The following assertions are equivalent and if either holds, then an almost complex structure J is said to be *integrable*.

   (a) There exists a holomorphic structure on M so that J is the associated almost complex structure.

   (b) The Nijenhuis tensor N_J vanishes.

Proof. Let $(w^1, \ldots, w^m)$ where $w^i = u^i + \sqrt{-1} v^i$ be another system of holomorphic coordinates on M. Let

$$J_1 : \partial_{x^i} \to \partial_{y^i}, \ J_1 : \partial_{y^i} \to -\partial_{x^i}, \ J_2 : \partial_{u^i} \to \partial_{v^i}, \text{ and } J_2 : \partial_{v^i} \to -\partial_{u^i}.$$

We must show $J_1 = J_2$. We use the Cauchy–Riemann equations to compute:

$$\partial_{x^i} = \partial_{x^i} u^j \cdot \partial_{u^j} + \partial_{x^i} v^j \cdot \partial_{v^j},$$
$$\partial_{y^i} = \partial_{y^i} u^j \cdot \partial_{u^j} + \partial_{y^i} v^j \cdot \partial_{v^j} = -\partial_{x^i} v^j \cdot \partial_{u^j} + \partial_{x^i} u^j \cdot \partial_{v^j},$$
$$J_2 \partial_{x^i} = \partial_u^i u^j \cdot \partial_{v^j} - \partial_{x^i} v^j \cdot \partial_{u^j} = \partial_{y^i} = J_1 \partial_{x^i},$$
$$J_2 \partial_{y^i} = -\partial_{x^i} v^j \cdot \partial_{v^j} - \partial_{x^i} u^j \cdot \partial_{u^j} = -\partial_{x^i} = J_1 \partial_{y^i}.$$

The proof that (2-a) implies (2-b) is a direct computation. The proof that (2-b) implies (2-a) is at the heart of the matter. It is a hard theorem in nonlinear partial differential equations that is beyond the scope of our current investigation and we refer to Newlander and Nirenberg [52] for details.   □

18.4.5 THE BUNDLES $\Lambda^{p,q} M$.

Let (M, J) be a holomorphic manifold of complex dimension $\mathfrak{m}$. Let $\vec{z} = (z^1, \ldots, z^m)$ be local holomorphic coordinates on M. We complexify the tangent and cotangent bundles. In the one-variable setting, Lemma 18.27 yields

$$f' = \tfrac{1}{2}\{(u_x + \sqrt{-1} u_x) - \sqrt{-1}(u_y + \sqrt{-1} v_y)\} \quad \text{if} \quad f \in \operatorname{Hol}(\mathcal{O}).$$

Consequently, we define $\partial_z := \frac{1}{2}\{\partial_x - \sqrt{-1}\partial_y\}$ and similarly $\partial_{\bar{z}} := \frac{1}{2}\{\partial_x + \sqrt{-1}\partial_y\}$. A function $f : \mathcal{O} \to \mathbb{C}$ is holomorphic if and only if $\partial_{\bar{z}} f = 0$ and in this setting $f' = \partial_z f$. In the multi-variable setting, we define:

$$dz^j := dx^j + \sqrt{-1}dy^j, \quad \partial_{z^j} := \frac{1}{2}(\partial_{x^j} - \sqrt{-1}\partial_{y^j}),$$
$$d\bar{z}^j := dx^j - \sqrt{-1}dy^j, \quad \partial_{\bar{z}^j} := \frac{1}{2}(\partial_{x^j} + \sqrt{-1}\partial_{y^j}).$$

Let $\langle \cdot, \cdot \rangle$ be the natural pairing between a tangent and a cotangent vector. We have

$$\langle \partial_{z^j}, dz^i \rangle = \delta^i_j, \quad \langle \partial_{z^j}, d\bar{z}^i \rangle = 0, \quad \langle \partial_{\bar{z}^j}, dz^i \rangle = 0, \quad \langle \partial_{\bar{z}^j}, d\bar{z}^i \rangle = \delta^i_j.$$

Let J^* be the dual endomorphism of T^*M so $\langle JX, \omega \rangle = \langle X, J^*\omega \rangle$. Then

$$J\partial_{x^j} = \partial_{y^j}, \qquad J\partial_{y^j} = -\partial_{x^j}, \qquad J^*dx^j = -dy^j, \qquad J^*dy^j = dx^j,$$
$$J\partial_{z^j} = \sqrt{-1}\partial_{z^j}, \quad J\partial_{\bar{z}^j} = -\sqrt{-1}\partial_{\bar{z}^j}, \quad J^*dz^j = \sqrt{-1}dz^j, \quad J^*d\bar{z}^j = -\sqrt{-1}d\bar{z}^j.$$

Given multi-indices $I = \{1 \leq i_1 < \cdots < i_p \leq \mathfrak{m}\}$ and $J = \{1 \leq j_1 < \cdots < j_q \leq \mathfrak{m}\}$, let

$$dz^I := dz^{i_1} \wedge \cdots \wedge dz^{i_p} \quad \text{and} \quad d\bar{z}^J := d\bar{z}^{j_1} \wedge \cdots \wedge d\bar{z}^{j_q}.$$

We may decompose $\Lambda^n M \otimes \mathbb{C} = \oplus_{p+q=n} \Lambda^{p,q} M$ where

$$\Lambda^{p,q} M := \mathrm{span}_{|I|=p, |J|=q}\{dz^I \wedge d\bar{z}^J\};$$

To simplify the notation, we omit the dependence on the complex structure. Suppose that $\vec{w} = (w^1, \ldots, w^{\mathfrak{m}})$ is another system of local holomorphic coordinates on M. Since w^i is holomorphic, the Cauchy–Riemann equations yield $\partial_{\bar{z}^i} w^j = 0$. Consequently, we have

$$dw^i = \partial_{z^j} w^i \cdot dz^j \quad \text{and} \quad d\bar{w}^i = \partial_{\bar{z}^j} \bar{w}^i \cdot d\bar{z}^j.$$

This shows that the bundles $\Lambda^{p,q} M$ are well defined and do not depend on the choice of local holomorphic coordinates. We extend J^* to the exterior algebra and obtain

$$J^* = (\sqrt{-1})^{p-q} \,\mathrm{Id} \quad \text{on} \quad \Lambda^{p,q} M. \tag{18.4.a}$$

A Riemannian metric g on M is said to be a *Hermitian metric* if $J^*g = g$, i.e., if

$$g(X, Y) = g(JX, JY)$$

for all tangent vectors X and Y. If g is an arbitrary Riemannian metric on M, then we can average over the action of J to define a Hermitian metric $\frac{1}{2}(g + J^*g)$. Thus, Hermitian metrics always exist and we assume g Hermitian henceforth. The triple (M, g, J) is said to be a *Hermitian manifold*. Let g be a Hermitian metric. We extend g to the complexified tangent bundle and exterior algebra to be conjugate linear in the second factor to define Hermitian inner products on the bundles $\Lambda^{p,q} M$. We obtain an orthogonal direct sum decomposition of the exterior algebra as a doubly graded algebra in the form

$$\Lambda^n(M) \otimes_{\mathbb{R}} \mathbb{C} = \oplus_{p+q=n} \Lambda^{p,q} M.$$

18.4.6 THE DOLBEAULT COMPLEX. If f is a smooth function on a holomorphic manifold (M, J), let

$$\partial f := (\partial_{z^j} f) dz^j \quad \text{and} \quad \bar{\partial} f := (\partial_{\bar{z}^j} f) d\bar{z}^j \,.$$

The operators $\partial : C^\infty(M) \to C^\infty(\Lambda^{1,0}M)$ and $\bar{\partial} : C^\infty(M) \to C^\infty(\Lambda^{0,1}M)$ are invariantly defined. A function f is holomorphic if and only if $\bar{\partial} f = 0$. Extend ∂ and $\bar{\partial}$ to the full exterior algebra by defining:

$$\partial(f_{I,J} dz^I \wedge d\bar{z}^J) := \partial f_{I,J} \wedge dz^I \wedge d\bar{z}^J \quad \text{and} \quad \bar{\partial}(f_{I,J} dz^I \wedge d\bar{z}^J) := \bar{\partial} f_{I,J} \wedge dz^I \wedge d\bar{z}^J \,.$$

We then have invariantly defined first-order partial differential operators

$$\partial : C^\infty(\Lambda^{p,q}M) \to C^\infty(\Lambda^{p+1,q}M) \quad \text{and} \quad \bar{\partial} : C^\infty(\Lambda^{p,q}M) \to C^\infty(\Lambda^{p,q+1}M)$$

which satisfy the relations

$$d = \partial + \bar{\partial} \quad \text{so} \quad \partial^2 = \bar{\partial}^2 = \partial\bar{\partial} + \bar{\partial}\partial = 0 \,. \tag{18.4.b}$$

A complex vector bundle over a holomorphic manifold M is said to be a *holomorphic vector bundle* if there exists a distinguished family of local frames $\{\vec{s}_\alpha\}$, called *holomorphic frames*, so the transition functions $\phi_{\alpha\beta}$ expressing $\vec{s}_\alpha = \phi_\alpha^\beta \vec{s}_\beta$ are holmorphic. A local section s to E is said to be a *holomorphic section* if the coefficient functions are holomorphic when s is written in terms of a local holomorphic frame. In this setting, define

$$\bar{\partial}_E(fs \otimes d\bar{z}^I) = s \otimes \bar{\partial} f \wedge d\bar{z}^I \,. \tag{18.4.c}$$

We then have an invariantly defined first-order partial differential operator

$$\bar{\partial}_E : C^\infty(\Lambda^{p,q}M \otimes E) \to C^\infty(\Lambda^{p,q+1}M \otimes E) \quad \text{with} \quad \bar{\partial}_E^2 = 0 \,.$$

Define the *Dolbeault cohomology groups with coefficients in E* by setting

$$H^{p,q}_{\mathrm{Dol}}(M, E) := \frac{\ker\{\bar{\partial}_E : C^\infty(E \otimes \Lambda^{p,q}M) \to C^\infty(E \otimes \Lambda^{p,q+1}M)\}}{\mathrm{Image}\{\bar{\partial}_E : C^\infty(E \otimes \Lambda^{p,q-1}M) \to C^\infty(E \otimes \Lambda^{p,q}M)\}} \,;$$

$\Lambda^{p,0}M$ is a holomorphic vector bundle and $H^{p,q}_{\mathrm{Dol}}(M, E) = H^{0,q}_{\mathrm{Dol}}(M, \Lambda^{p,0}M \otimes E)$. Thus, it is customary to take $p = 0$ and define

$$H^q_{\mathrm{Dol}}(M, E) := H^{0,q}_{\mathrm{Dol}}(M, E) \,.$$

Define the *Dolbeault complex with coefficients in E* by:

$$\mathcal{E}_{\mathrm{Dol}}(M, E) := \{\sqrt{2}\bar{\partial}_E : C^\infty(\Lambda^{0,q}M \otimes E) \to C^\infty(\Lambda^{0,q+1}M \otimes E)\} \,. \tag{18.4.d}$$

Let (E, h) be a holomorphic Hermitian vector bundle over a Hermitian manifold (M, g, J). Let δ''_E be the formal adjoint of $\bar{\partial}_E$. The operator $\bar{\partial}_E$ only depends on the holomorphic structures; the adjoint δ''_E depends on the choice of g and h. Notice that at this stage we have not defined the operators ∂_E and δ'_E as we do not need them yet; we postpone that task until Section 18.5.4 when we discuss the formula of Nakano. Define the associated *normalized complex Laplacian* by setting

$$\Delta^{p,q}_E := 2\{\delta''_\mathcal{E}\bar{\partial}_E + \bar{\partial}_E\delta''_\mathcal{E}\} : C^\infty(\Lambda^{p,q}M \otimes E) \to C^\infty(\Lambda^{p,q}M \otimes E). \tag{18.4.e}$$

The factor of $\sqrt{2}$ in Equation (18.4.d) and the factor of 2 in Equation (18.4.e) are present to ensure that $\Delta^{p,q}_E$ is an operator of Laplace type; apart from that, they play no role. If $\mathcal{U}$ is an open subset of M, let $\mathbf{Hol}(E)$ be the *sheaf of holomorphic sections* to E over $\mathcal{U}$. The associated cohomology groups (see Section 16.1.5) are denoted by $H^q(M, \mathbf{Hol}(E))$. One has the following result of Dolbeault.

Theorem 18.31   Let (E, h) be a holomorphic Hermitian vector bundle over a Hermitian manifold (M, g, J).

1. $\mathcal{E}_{\mathrm{Dol}}(M, E)$ is an elliptic complex of Dirac type.
2. The map $\omega \to [\omega]$ is an isomorphism from $\ker\{\Delta^{0,q}_E\} \to H^q_{\mathrm{Dol}}(E)$.
3. $\dim(H^q_{\mathrm{Dol}}(E)) < \infty$.
4. $H^q(M, \mathbf{Hol}(E)) = H^q_{\mathrm{Dol}}(E)$.
5. $\mathbf{Hol}(E)(M) = H^0(M, \mathbf{Hol}(E))$ is the finite-dimensional vector space (possibly trivial) of global holomorphic sections to E.

Proof. Since $\bar{\partial}^2_E = 0$, the only question at issue in the proof of Assertion 1 is to show that $2\Delta_{p,q,E}$ is an operator of Laplace type. This is a question on the leading symbol and, consequently, it suffices to compute in flat space. In that setting, E plays no role and can be ignored. In flat space, $\bar{\partial} = \mathrm{ext}(d\bar{z}^i)\partial_{\bar{z}^i}$, $\delta'' = -\mathrm{int}(dz^j)\partial_{z^j}$, and

$$2\{\bar{\partial}\delta'' + \delta''\bar{\partial}\} = -2\{\mathrm{ext}(d\bar{z}^i)\,\mathrm{int}(dz^j) + \mathrm{int}(dz^j)\,\mathrm{ext}(d\bar{z}^i)\}\partial_{z^i}\partial_{\bar{z}^j}$$
$$= -2\{2\delta_{ij}\partial_{z^i}\partial_{z^j}\} = -\sum_i(\partial_{x^i} - \sqrt{-1}\partial_{y^i})(\partial_{x^i} + \sqrt{-1}\partial_{y^i})$$
$$= -\sum_i(\partial^2_{x^i} + \partial^2_{y^i}).$$

Assertions 2 and 3 follow from Assertion 1 and from Theorem 17.24. Assertions 4 and 5 follow from the Dolbeault Lemma (which we will establish subsequently as Theorem 19.7 in Section 19.1.6) in exactly the same fashion that the isomorphism

$$H^p(M, \mathbb{C}) = H^p_{\mathrm{deR}}(M)$$

followed from the Poincaré Lemma (see Theorem 8.14 of Book II). We shall omit the details in the interests of brevity as it is a diagram chase based on the long exact sequences arising from Equation (16.1.b) using Lemma 16.9.   □

18.4.7 THE CHERN CONNECTION.

Lemma 18.32   Let (M, g, J) be a Hermitian manifold and let (E, h) be a holomorphic Hermitian vector bundle over M.

1. Given a point P of M, there is a holomorphic frame $\vec{s}$ so $dh_{\vec{s},\alpha\bar{\beta}}(P) = 0$.

2. Let $\vec{s} = (s_1, \ldots, s_\ell)$ be a local holomorphic frame for V. Let $h_{\vec{s},\alpha\bar{\beta}} := h(s_\alpha, s_\beta)$, let $h^{\vec{s},\alpha\bar{\beta}}$ be the inverse matrix, and let $\omega_{\vec{s},h} = h_{\vec{s}}^{-1} \partial h_{\vec{s}}$ be the connection one-form of $\nabla_{\vec{s},h}$, i.e., we have that $\nabla_{\vec{s},h} s_\alpha = h^{\vec{s},\beta\bar{\gamma}} \partial h_{\vec{s},\alpha\bar{\gamma}} \otimes s_\beta$. Then $\nabla_{\vec{s},h}$ is independent of $\vec{s}$ and defines a unitary connection ∇_h on V called the *Chern connection*.

3. The associated curvature operator $\mathcal{R}$ belongs to $C^\infty(\Lambda^{1,1} M \otimes \mathrm{End}(V))$.

Proof. Let $\vec{s} = (s_1, \ldots, s_\ell)$ be a local holomorphic frame for V defined on an open subset $\mathcal{O}_{\vec{s}}$ of M. Let $\vec{t} = (t_1, \ldots, t_\ell)$ be another local holomorphic frame for V defined on $\mathcal{O}_{\vec{t}}$. On the common domain of definition $\mathcal{O}_{\vec{s}} \cap \mathcal{O}_{\vec{t}}$, we can express $t_\alpha = \phi_\alpha^u s_u$ where $\phi := (\phi_\alpha^u)$ is a holomorphic map from $\mathcal{O}_{\vec{s}} \cap \mathcal{O}_{\vec{t}}$ to $\mathrm{GL}(\ell, \mathbb{C})$. Let Φ_u^α be the inverse matrix so $s_u = \Phi_u^\alpha t_\alpha$. We compute

$$h_{\vec{t},\sigma\bar{\gamma}} = h(\phi_\sigma^u s_u, \phi_\gamma^v s_v) = \phi_\sigma^u h_{\vec{s},u\bar{v}} \bar{\phi}_\gamma^v .$$

To establish Assertion 1, we choose the local coordinates $\vec{z} = (z^1, \ldots, z^m)$ so that $\vec{z} = 0$ corresponds to the point P. We make a linear change of coordinates using the Gram–Schmidt process (see the discussion of Section 16.3.3) to assume $\vec{s}$ is chosen so $h_{\alpha\bar{\beta}}(0) = \delta_{\alpha\beta}$. Let

$$\phi_\alpha^u := \delta_\alpha^u - \{\partial_{z^\beta} h_{\alpha u}(0)\} z^\beta .$$

Since $\det\{(\phi_\alpha^u)\}(0) = 1$, ϕ takes values in $\mathrm{GL}(\ell, \mathbb{C})$ on some neighborhood of P. We compute:

$$h_{\vec{t},\sigma\bar{\gamma}} = (\delta_\sigma^u - \{\partial_{z^\beta} h_{\sigma u}(0)\} z^\beta) h_{\vec{s},u\bar{v}} (\delta_\gamma^v - \overline{\{\partial_{z^\tau} h_{\gamma v}(0)\} z^\tau}),$$

$$\{\partial_{z^\alpha} h_{\vec{t},\sigma\bar{\gamma}}\}(0) = -\{\partial_{z^\alpha} h_{\sigma u}(0)\} \delta_{u\bar{v}} \delta_\gamma^v + \delta_\sigma^u \{\partial_{z^\alpha} h_{\vec{s},u\bar{v}}(0)\} \delta_\gamma^v = 0 .$$

This shows $\partial h_{\vec{t},\sigma\bar{\gamma}}(0) = 0$. Since $\overline{h_{\vec{t},\sigma\bar{\gamma}}} = h_{\vec{t},\gamma\bar{\sigma}}$, we conclude as well $\bar{\partial} h_{\vec{t},\sigma\bar{\gamma}}(0) = 0$. This proves Assertion 1. We begin the proof of Assertion 2 by computing:

$$\omega_{\vec{t},\alpha}^\beta = h^{\vec{t},\beta\bar{\gamma}} \partial h_{\vec{t},\alpha\bar{\gamma}} = \bar{\Phi}_b^\gamma h^{\vec{s},a\bar{b}} \Phi_a^\beta \{\partial \phi_\alpha^u \cdot h_{\vec{s},u\bar{v}} + \phi_\alpha^u \partial h_{\vec{s},u\bar{v}}\} \bar{\phi}_\gamma^v$$

$$= \delta_b^v h^{\vec{s},a\bar{b}} \Phi_a^\beta \{\partial \phi_\alpha^u \cdot h_{\vec{s},u\bar{v}} + \phi_\alpha^u \partial h_{\vec{s},u\bar{v}}\} = h^{\vec{s},a\bar{b}} \Phi_a^\beta \{\partial \phi_\alpha^u \cdot h_{\vec{s},u\bar{b}} + \phi_\alpha^u \partial h_{\vec{s},u\bar{b}}\}$$

$$= \Phi_a^\beta \partial \phi_\alpha^a + \Phi_a^\beta \phi_\alpha^u \omega_{\vec{s},u}^a .$$

We may then show ∇ is invariantly defined by computing:

$$\nabla_{\vec{s}}(t_\alpha) = \nabla_{\vec{s}}(\phi_\alpha^a s_a) = \partial \phi_\alpha^a s_a + \phi_\alpha^u \omega_{\vec{s},u}^a s_a = \Phi_a^\beta (\partial \phi_\alpha^a + \phi_\alpha^u \omega_{\vec{s},u}^a) t_\beta = \nabla_{\vec{t}}(t_\alpha) .$$

To prove Assertion 3, we use Assertion 1 to choose a local holomorphic frame so $h_{\alpha\bar{\beta}}(P) = \delta_{\alpha\beta}$ and $dh(P) = 0$. Thus, $\nabla h(P) = 0$ and $R(P) = \bar{\partial} \partial h_s(P)$ is of type (1,1).   □

18.4.8 THE RIEMANN–ROCH THEOREM [65]. We say that (M, g) is a *Riemann surface* if M is an oriented two-dimensional manifold which is equipped with a Riemannian metric g. We may then define an almost complex structure by letting J be counterclockwise rotation through an angle of $\frac{\pi}{2}$. The Newlander–Nirenberg Theorem (Theorem 18.30) then shows J is integrable so (M, g, J) is a Hermitian manifold. In particular, we obtain *isothermal coordinates* to express $g = e^{\phi}(dx^2 + dy^2)$.

Theorem 18.33 Let (E, h) be a holomorphic Hermitian vector bundle over a compact Riemann surface (M, g). Then $\text{Index}(\mathcal{E}_{\text{Dol}}(M, E)) = \frac{1}{2}\chi(M) + \int_M c_1(E, h)$.

18.4.9 THE HIRZEBRUCH–RIEMANN–ROCH THEOREM [38]. We use the Chern connection ∇_h to define the Chern character of E (see Section 16.1.6.c). Let (M, g, J) be a Hermitian manifold. We use J to regard TM as a complex vector bundle and to define the *Todd genus* $\text{Td}(M, g, J)$ (see Section 16.1.6.d). Theorem 18.33 generalizes to become the following result.

Theorem 18.34 Let (E, h) be a holomorphic Hermitian vector bundle over a compact Hermitian manifold (M, g, J). Then

$$\text{Index}(\mathcal{E}_{\text{Dol}}(M, E)) = \int_M \{\text{Td}(M, g, J) \wedge \text{ch}(E, h)\}_m.$$

18.5 DUALITY AND VANISHING THEOREMS IN COMPLEX GEOMETRY

We discuss two versions of the Serre Duality Theorem. Let (E, h) be a holomorphic Hermitian vector bundle over a compact Hermitian manifold (M, g, J). Poincaré duality extends to this setting; in Section 18.5.1, we show that:

$$\dim(H^{p,q}(M, E)) = \dim(H^{m-p, m-q}(M, E^*)).$$

The next duality theorem is only valid in the Kähler setting. In Section 18.5.2, we introduce some basic facts concerning Kähler geometry. With this material in hand, in Section 18.5.3, we show that if (M, g, J) is Kähler, then

$$H^n_{\text{deR}}(M) \otimes_{\mathbb{R}} \mathbb{C} = \oplus_{p,q=n} H^{p,q}(M) \quad \text{and} \quad \dim(H^{p,q}(M)) = \dim(H^{q,p}(M)).$$

In Section 18.5.4, we establish the formula of Nakano and the Bochner–Kodaira–Nakano identity. We conclude in Section 18.5.5 by proving the Kodaira–Akizuki–Nakano Vanishing Theorem. This will play a central role in our discussion of the Kodaira Embedding Theorem subsequently in Chapter 19. It is the analogue of the Bochner Vanishing Theorem (Theorem 18.11) and the Lichnerowicz Vanishing Theorem (Theorem 18.23) in the complex category although considerably more difficult to establish.

18.5.1 SERRE DUALITY FOR HERMITIAN MANIFOLDS.

If $\mathcal{O}$ is an open subset of $\mathbb{C}^m$ and if $\vec{F} : \mathcal{O} \to \mathbb{C}^m$ is holomorphic, we showed that $\det(\vec{F}_r') = |\det(\vec{F}_c')|^2$ in the proof of Theorem 18.28. Consequently, any holomorphic manifold is orientable. We normalize the sign by requiring that in any local holomorphic coordinate system, the orientation is defined by $dx^1 \wedge dy^1 \wedge \cdots \wedge dx^m \wedge dy^m$. Let g be a Hermitian metric and let orn be the associated unit volume form. We begin by examining how Clifford multiplication by orn acts on the spaces $\Lambda^{p,q} M$.

Lemma 18.35 Let (M, g, J) be a Hermitian manifold of complex dimension m.

1. $\partial = \mathrm{ext}(dz^i)\nabla_{\partial_{z^i}}$, $\bar{\partial} = \mathrm{ext}(d\bar{z}^i)\nabla_{\partial_{\bar{z}^i}}$, $\delta'' = -\mathrm{int}(dz^i)\nabla_{\partial_{z^i}}$, and $\delta' = -\mathrm{int}(d\bar{z}^i)\nabla_{\partial_{\bar{z}^i}}$.

2. $c_\Lambda(\mathrm{orn}) : \Lambda^{p,q} M \to \Lambda^{m-q,m-p} M$.

Proof. It is an easy algebraic exercise to verify that

$$\mathrm{ext} : \Lambda^{1,0} M \otimes \Lambda^{p,q} M \to \Lambda^{p+1,q} M, \quad \mathrm{ext} : \Lambda^{0,1} M \otimes \Lambda^{p,q} M \to \Lambda^{p,q+1} M,$$
$$\mathrm{int} : \Lambda^{1,0} M \otimes \Lambda^{p,q} M \to \Lambda^{p,q-1} M, \quad \mathrm{int} : \Lambda^{0,1} M \otimes \Lambda^{p,q} M \to \Lambda^{p-1,q} M . \tag{18.5.a}$$

We apply Lemma 16.4 to express $d = \mathrm{ext} \circ \nabla$ and $\delta = -\mathrm{int} \circ \nabla$. Assertion 1 follows from Equation (18.5.a) and from the following computation:

$$\mathrm{ext}(dz^i)\nabla_{\partial_{z^i}} + \mathrm{ext}(d\bar{z}^i)\nabla_{\partial_{\bar{z}^i}}$$
$$= \tfrac{1}{2}\mathrm{ext}(dx^i + \sqrt{-1}dy^i)(\nabla_{\partial_{x^i}} - \sqrt{-1}\nabla_{\partial_{y^i}})$$
$$\quad + \tfrac{1}{2}\mathrm{ext}(dx^i - \sqrt{-1}dy^i)(\nabla_{\partial_{x^i}} + \sqrt{-1}\nabla_{\partial_{y^i}})$$
$$= \mathrm{ext}(dx^i)\nabla_{\partial_{x^i}} + \mathrm{ext}(dy^i)\nabla_{\partial_{y^i}} = d,$$

$$\mathrm{int}(dz^i)\nabla_{\partial_{z^i}} + \mathrm{int}(d\bar{z}^i)\nabla_{\partial_{\bar{z}^i}}$$
$$= -\tfrac{1}{2}\mathrm{int}(dx^i + \sqrt{-1}dy^i)(\nabla_{\partial_{x^i}} - \sqrt{-1}\nabla_{\partial_{y^i}})$$
$$\quad - \tfrac{1}{2}\mathrm{int}(dx^i - \sqrt{-1}dy^i)(\nabla_{\partial_{x^i}} + \sqrt{-1}\nabla_{\partial_{y^i}})$$
$$= -\mathrm{int}(dx^i)\nabla_{\partial_{x^i}} - \mathrm{int}(dy^i)\nabla_{\partial_{y^i}} = \delta .$$

Suppose first $m = 1$ in the proof of Assertion 2. This is a purely algebraic computation so we can assume $M = \mathbb{C}$ with the flat Euclidean metric and usual holomorphic coordinate $z = x + \sqrt{-1}y$; $c_\Lambda(\mathrm{orn}) = c_\Lambda(dx)c_\Lambda(dy)$. One establishes the desired result in this special case by computing

$$c_\Lambda(\mathrm{orn})\mathbb{1} = c_\Lambda(dx)c_\Lambda(dy)\mathbb{1} = dx \wedge dy = \tfrac{1}{2}\sqrt{-1}dz \wedge d\bar{z},$$
$$c_\Lambda(\mathrm{orn})\left\{\tfrac{1}{2}\sqrt{-1}dz \wedge d\bar{z}\right\} = c_\Lambda(dx)c_\Lambda(dy)dx \wedge dy = -\mathbb{1},$$
$$c_\Lambda(\mathrm{orn})\{dz\} = c_\Lambda(dx)c_\Lambda(dy)(dx + \sqrt{-1}dy) = dy - \sqrt{-1}dx = -\sqrt{-1}dz,$$
$$c_\Lambda(\mathrm{orn})\{d\bar{z}\} = c_\Lambda(dx)c_\Lambda(dy)(dx - \sqrt{-1}dy) = dy + \sqrt{-1}dx = \sqrt{-1}d\bar{z},$$

$$c_\Lambda(\mathrm{orn}) : \Lambda^{0,0}\mathbb{C} \to \Lambda^{1,1}\mathbb{C}, \qquad c_\Lambda(\mathrm{orn}) : \Lambda^{1,1}\mathbb{C} \to \Lambda^{0,0}\mathbb{C},$$
$$c_\Lambda(\mathrm{orn}) : \Lambda^{1,0}\mathbb{C} \to \Lambda^{1,0}\mathbb{C}, \qquad c_\Lambda(\mathrm{orn}) : \Lambda^{0,1}\mathbb{C} \to \Lambda^{0,1}\mathbb{C}.$$

The desired result if $m > 1$ follows from the case $m = 1$ using the decomposition

$$\Lambda\mathbb{C}^m = \Lambda\mathbb{C} \otimes \cdots \otimes \Lambda\mathbb{C} \quad \text{and} \quad c_\Lambda\mathrm{orn}_m = c_\Lambda\mathrm{orn}_1 \otimes \cdots \otimes c_\Lambda\mathrm{orn}_1. \qquad \square$$

Theorem 18.36   Let (E, h) be a holomorphic Hermitian vector bundle over a compact Hermitian manifold (M, g, J). Let $\Theta_h : E \to E^*$ be the conjugate linear map which is defined by setting $\Theta_h(e_1)\{e_2\} = h(e_2, e_1)$. Define $\check{c}_\Lambda(\mathrm{orn})\eta = \overline{c_\Lambda(\mathrm{orn})\eta} = c_\Lambda(\mathrm{orn})\bar\eta$ for $\eta \in \Lambda_\mathbb{C}(M)$.

1. $\check{c}_\Lambda(\mathrm{orn}) \otimes \Theta_h : \Lambda^{p,q} M \otimes E \to \Lambda^{m-p,m-q} M \otimes E^*$ is a conjugate linear isometry.
2. $\{\check{c}_\Lambda(\mathrm{orn}) \otimes \Theta_h\}^2 = (-1)^m \mathrm{Id}$.
3. $(\check{c}_\Lambda(\mathrm{orn}) \otimes \Theta_h)(\bar\partial_E + \delta''_E) = -(\bar\partial_E + \delta''_E)(\check{c}_\Lambda(\mathrm{orn}) \otimes \Theta_h)$.
4. $\check{c}_\Lambda(\mathrm{orn}) \otimes \Theta_h : H^{p,q}(M, E) \to H^{m-p,m-q}(M, E^\star)$ is an isomorphism.

Proof. Assertions 1 and 2 are immediate from the definition. To prove Assertion 3, suppose for the moment $E = \mathbb{1}$ is the trivial line bundle with flat inner product h. By Lemma 18.35, $c_\Lambda(\mathrm{orn}) : \Lambda^{p,q} \to \Lambda^{m-q,m-p}$ so

$$\check{c}_\Lambda(\mathrm{orn}) : \Lambda^{p,q} M \to \Lambda^{m-p,m-q} M,$$
$$\check{c}_\Lambda(\mathrm{orn})\partial : C^\infty(\Lambda^{p,q} M) \to C^\infty(\Lambda^{m-p-1,m-q} M),$$
$$\check{c}_\Lambda(\mathrm{orn})\bar\partial : C^\infty(\Lambda^{p,q} M) \to C^\infty(\Lambda^{m-p,m-q-1} M),$$
$$\delta'\check{c}_\Lambda(\mathrm{orn}) : C^\infty(\Lambda^{p,q} M) \to C^\infty(\Lambda^{m-p-1,m-q} M),$$
$$\delta''\check{c}_\Lambda(\mathrm{orn}) : C^\infty(\Lambda^{p,q} M) \to C^\infty(\Lambda^{m-p,m-q-1} M).$$

By Theorem 18.12, $c_\Lambda(\mathrm{orn}) \circ d = (-1)^{2m-1}\delta \circ c_\Lambda(\mathrm{orn})$. Consequently, by examining the bi-grading, we have $\check{c}_\Lambda(\mathrm{orn})\bar\partial = -\delta''\check{c}_\Lambda(\mathrm{orn})$. Since $\check{c}_\Lambda(\mathrm{orn})^2 = (-1)^m \mathrm{Id}$, we also have $\bar\partial\check{c}_\Lambda(\mathrm{orn}) = -\check{c}_\Lambda(\mathrm{orn})\delta''$ and thus $\check{c}_\Lambda(\mathrm{orn})(\bar\partial + \delta'') = -(\bar\partial + \delta'')\check{c}_\Lambda(\mathrm{orn})$. In the general case, set $\mathcal{E}(h) := (\check{c}_\Lambda(\mathrm{orn}) \otimes \Theta_h)(\bar\partial_E + \delta''_E) + (\bar\partial_E + \delta''_E)(\check{c}_\Lambda(\mathrm{orn}) \otimes \Theta_h)$. The error $\mathcal{E}(h)$ is an invariantly defined endomorphism which is linear in the first derivatives of the auxiliary Hermitian inner product on E. By Lemma 18.32, we may choose a holomorphic frame such that the derivatives of h vanish at the point. This shows that $\mathcal{E}(h) = 0$ at this point and hence vanishes identically. Assertion 3 now follows. Assertion 4 follows from Theorem 18.31 and from Assertion 3. $\qquad \square$

18.5.2 KÄHLER METRICS.
A good reference for this material is Ballman [8]. If (M, g, J) is a Hermitian manifold, let

$$\Omega(X, Y) := g(JX, Y) = -g(X, JY).$$

We have $\Omega(Y, X) = g(JY, X) = g(X, JY) = -\Omega(X, Y)$ so Ω is a skew-symmetric bilinear form on the tangent space and hence a real two-form called the *Kähler form*. Since

$$\Omega(JX, JY) = \Omega(X, Y),$$

Ω is a form of type $(1, 1)$. Let $z^i = x^i + \sqrt{-1}y^i$ be the usual holomorphic coordinates on $\mathbb{C}^m$. Let ds_e^2 be the Euclidean metric and Ω_e the corresponding Kähler form. Then

$$ds_e^2 = \sum_{i=1}^m \{dx_i^2 + dy_i^2\} \quad \text{and} \quad \Omega_e = \sum_{i=1}^m dx^i \wedge dy^i = \tfrac{1}{2}\sqrt{-1}\sum_{i=1}^m dz^i \wedge d\bar{z}^i. \qquad (18.5.\text{b})$$

Let $\mathcal{R}$ be the curvature operator and let R be the curvature tensor of the Levi–Civita connection. We recall the following result from Book II (see Theorem 4.9).

Theorem 18.37 Let (M, g, J) be a Hermitian manifold.

1. The following conditions are equivalent and if any is satisfied, then (M, g, J) is said to be a Kähler manifold:

   (a) $d\Omega = 0$.

   (b) $\nabla\Omega = 0$.

   (c) $\nabla J = 0$.

   (d) There are local holomorphic coordinates centered at any point P of M so all the first derivatives of the metric tensor g_{ij} vanish at P.

2. If (M, g, J) is a Kähler manifold, then
$$J\mathcal{R}(X, Y) = \mathcal{R}(X, Y)J \quad \text{and} \quad R(JX, JY, Z, W) = R(X, Y, Z, W).$$

The Kähler condition is stronger than complex integrability; Example 4.4.1 of Book II shows $S^1 \times S^3$ admits a natural holomorphic structure but no Kähler metric.

18.5.3 SERRE DUALITY IN THE KÄHLER SETTING. We have by Lemma 16.4 that $d = \text{ext} \circ \nabla$ and $\delta = -\text{int} \circ \nabla$. Our first task is to establish a similar representation in the Kähler setting. Let δ' be the formal adjoint of ∂ and let δ'' be the formal adjoint of $\bar{\partial}$. Let $\text{ext}(\Omega)$ be exterior multiplication by the Kähler form and $\text{int}(\Omega)$ be the dual endomorphism of the cotangent bundle, interior multiplication by the Kähler form.

Lemma 18.38 Let (M, g, J) be a Kähler manifold. Then
1. $\bar{\partial}\,\text{int}(\Omega) - \text{int}(\Omega)\bar{\partial} = \sqrt{-1}\delta'$ and $\partial\,\text{int}(\Omega) - \text{int}(\Omega)\partial = -\sqrt{-1}\delta''$.
2. If ω is an n-form, then $\{\text{ext}(\Omega)\,\text{int}(\Omega) - \text{int}(\Omega)\,\text{ext}(\Omega)\}\,\omega = (n - m)\omega$.

Proof. Set $\mathcal{E} := \bar{\partial}\,\text{int}(\Omega) - \text{int}(\Omega)\bar{\partial} - \sqrt{-1}\delta'$. If we can show that $\mathcal{E}$ vanishes in flat space, it will then follow that $\mathcal{E}$ is an invariantly defined endomorphism which is linear in the one-jets of the metric. Fix a point P. By Theorem 18.37, we can always choose holomorphic coordinates so $g_{ij}(P) = \delta_{ij} + O(\|\bar{z}\|^2)$. It then follows that the error $\mathcal{E}$ vanishes at P. Since $\mathcal{E}$ is invariantly defined, $\mathcal{E}$ will vanish and we will obtain $\bar{\partial}\,\text{int}(\Omega) - \text{int}(\Omega)\bar{\partial} - \sqrt{-1}\delta' = 0$ in general; the corresponding identity for ∂ will then follow by taking the complex conjugate. Thus, we may work in flat space. By Equation (18.5.b),

$$\text{ext}(\Omega) = \sum_j \text{ext}(dx^j)\,\text{ext}(dy^j) \quad \text{and} \quad \text{int}(\Omega) = \sum_j \text{int}(dy^j)\,\text{int}(dx^j)\,.$$

Fix i and let $\mathcal{E}_i := \left[\,\text{ext}(d\bar{z}^i), \sum_j \text{int}(dy^j)\,\text{int}(dx^j)\right]$. By Lemma 18.35, $\bar{\partial} = \text{ext}(d\bar{z}^i)\partial_{\bar{z}^i}$ and $\delta' = -\text{int}(d\bar{z}^i)\partial_{\bar{z}^i}$. Consequently,

$$\bar{\partial}\,\text{int}(\Omega) - \text{int}(\Omega)\bar{\partial} - \sqrt{-1}\delta' = \sum_i (\mathcal{E}_i + \sqrt{-1}\,\text{int}(d\bar{z}^i))\partial_{\bar{z}^i}\,.$$

Thus, we must show $\mathcal{E}_i = -\sqrt{-1}\,\text{int}(d\bar{z}^i)$. Because $\text{ext}(\eta)\,\text{int}(\xi) + \text{int}(\xi)\,\text{ext}(\eta) = g(\xi, \eta)$, we may set $i = j$ in the sum defining $\mathcal{E}_i$. We establish Assertion 1 by computing:

$$\begin{aligned}
\mathcal{E}_i &= \text{ext}(dx^i - \sqrt{-1}dy^i)\,\text{int}(dy^i)\,\text{int}(dx^i) \\
&\quad - \text{int}(dy^i)\,\text{int}(dx^i)\,\text{ext}(dx^i - \sqrt{-1}dy^i) \\
&= -\text{int}(dy^i)\{\text{ext}(dx^i)\,\text{int}(dx^i) + \text{int}(dx^i)\,\text{ext}(dx^i)\} \\
&\quad - \sqrt{-1}\{\text{ext}(dy^i)\,\text{int}(dy^i) + \text{int}(dy^i)\,\text{ext}(dy^i)\}\,\text{int}(dx^i) \\
&= -\text{int}(dy^i) - \sqrt{-1}\,\text{int}(dx^i) = -\sqrt{-1}\,\text{int}(d\bar{z}^i)\,.
\end{aligned}$$

Since Assertion 2 is a purely algebraic identity, we can work in flat space and assume the manifold in question is $\mathbb{C}^m$ with the usual flat structures. Suppose first $m = 1$. We show Assertion 2 holds in this special case by computing:

$$\begin{aligned}
(\text{ext}(\Omega)\,\text{int}(\Omega) - \text{int}(\Omega)\,\text{ext}(\Omega)(dx \wedge dy)) &= dx \wedge dy, \\
(\text{ext}(\Omega)\,\text{int}(\Omega) - \text{int}(\Omega)\,\text{ext}(\Omega))1 &= -1, \\
(\text{ext}(\Omega)\,\text{int}(\Omega) - \text{int}(\Omega)\,\text{ext}(\Omega)(dx \wedge dy))dx &= 0, \\
(\text{ext}(\Omega)\,\text{int}(\Omega) - \text{int}(\Omega)\,\text{ext}(\Omega)(dx \wedge dy))dy &= 0\,.
\end{aligned}$$

More generally, decompose $\mathbb{C}^m = \mathbb{C} \oplus \mathbb{C}^{m-1}$, $\Lambda\mathbb{C}^m = \Lambda\mathbb{C} \otimes_{\mathbb{C}} \Lambda\mathbb{C}^{m-1}$, and $\omega = \omega_1 \wedge \omega_2$ where $\omega_1 \in \Lambda^{n_1}\mathbb{C}$ (for $n_1 = 0, 1$) and $\omega_2 \in \Lambda\mathbb{C}^{m-1}$. Let $\Omega_1 = dx^1 \wedge dy^1 \in \Lambda^2\mathbb{C}$. We have

$$\begin{aligned}
\{\text{ext}(\Omega_1)\,\text{int}(\Omega_1) &- \text{int}(\Omega_1)\,\text{ext}(\Omega_1)\}(w_1 \wedge w_2) \\
&= (\{\text{ext}(\Omega_1)\,\text{int}(\Omega_1) - \text{int}(\Omega_1)\,\text{ext}(\Omega_1)\}w_1) \wedge w_2 \qquad (18.5.\text{c}) \\
&= (n_1 - 1)\omega_1 \wedge \omega_2\,.
\end{aligned}$$

Let $\Omega_i := dx^i \wedge dy^i$. Because $\text{ext}(\Omega_i)$ commutes with $\text{int}(\Omega_j)$ for $i \neq j$, we have

$$[\text{ext}(\Omega), \text{int}(\Omega)] = \sum_{i,j}[\text{ext}(\Omega_i), \text{int}(\Omega_j)] = \sum_i[\text{ext}(\Omega_i), \text{int}(\Omega_i)]\,.$$

We may therefore sum Equation (18.5.c) to establish Assertion 2. $\qquad\square$

We now establish a second duality isomorphism in the Kähler setting which is also called *Serre duality*.

Theorem 18.39   Let (M, g, J) be a compact Kähler manifold. Then

1. $d\delta + \delta d = 2(\partial\delta' + \delta'\partial) = 2(\bar\partial\delta'' + \delta''\bar\partial)$.
2. $H^n_{\text{deR}}(M) \otimes_\mathbb{R} \mathbb{C} = \oplus_{p,q=n} H^{p,q}(M)$.
3. Complex conjugation mapping $H^{p,q}(M)$ to $H^{q,p}(M)$ is an isomorphism.

Proof. We use Lemmas 18.35 and 18.38 to compute:

$$\sqrt{-1}(\partial\delta' + \delta'\partial) = \partial\bar\partial \operatorname{int}(\Omega) - \partial \operatorname{int}(\Omega)\bar\partial + \bar\partial \operatorname{int}(\Omega)\partial - \operatorname{int}(\Omega)\bar\partial\partial,$$
$$-\sqrt{-1}(\partial\delta'' + \delta''\bar\partial) = \bar\partial\partial \operatorname{int}(\Omega) - \bar\partial \operatorname{int}(\Omega)\partial + \partial \operatorname{int}(\Omega)\bar\partial - \operatorname{int}(\Omega)\partial\bar\partial \, .$$

Since $\bar\partial\partial + \partial\bar\partial = 0$, it follows that $\partial\delta' + \delta'\partial = \bar\partial\delta'' + \delta''\bar\partial$. We compute:

$$\begin{aligned}\bar\partial\delta' + \delta'\bar\partial &= -\sqrt{-1}\{\bar\partial(\bar\partial \operatorname{int}(\Omega) - \operatorname{int}(\Omega)\bar\partial) + (\bar\partial \operatorname{int}(\Omega) - \operatorname{int}(\Omega)\bar\partial)\bar\partial\} \\ &= -\sqrt{-1}\{-\bar\partial \operatorname{int}(\Omega)\bar\partial + \bar\partial \operatorname{int}(\Omega)\bar\partial\} = 0 \, .\end{aligned}$$

Dually we have $\partial\delta'' + \delta''\partial = 0$. We complete the proof of Assertion 1 by computing:

$$\begin{aligned}d\delta + \delta d &= (\partial + \bar\partial)(\delta' + \delta'') + (\delta' + \delta'')(\partial + \bar\partial) \\ &= \partial\delta' + \bar\partial\delta' + \partial\delta'' + \bar\partial\delta'' + \delta'\partial + \delta'\bar\partial + \delta''\partial + \delta''\bar\partial \\ &= 2(\partial\delta' + \delta'\partial) \, .\end{aligned}$$

We use the Hodge Decomposition Theorem to identify the cohomology group $H^{p,q}(M)$ with $\ker\{\bar\partial\delta'' + \delta''\bar\partial\}$. This in turn can be identified, using Assertion 1, with $\ker\{d\delta + \delta d\}$. Summing over $p + q = n$ and using the Hodge Decomposition Theorem applied to d then identifies $\oplus_{p+q=n} H^{p,q}(M)$ with $H^n_{\text{deR}}(M)$. Assertion 2 follows. Assertion 3 is then immediate.   □

By combining Theorems 18.39 and 18.36, we obtain

$$\dim(H^{p,q}(M)) = \dim(H^{q,p}(M)) = \dim(H^{m-p,m-q}(M)) = \dim(H^{m-q,m-p}(M)) \, .$$

18.5.4 THE BOCHNER–KODAIRA–NAKANO IDENTITY[51]. This is an analogue of the Weitzenböck identity for Hermitian manifolds. Let (L, h) be a holomorphic Hermitian line bundle over a Hermitian manifold (M, g, J). If s is a local non-vanishing holomorphic section to L, set $h_s = h(s, s)$. By Lemma 18.32, the connection one-form of the Chern connection ∇_h is $\omega_s = h_s^{-1}\partial h_s = \partial \log h_s$ and, consequently, the curvature is $d\omega_s = d\partial \log h_s = \bar\partial\partial \log h_s$. Therefore, the first Chern form is given by

$$c_1(L, h) = \frac{\sqrt{-1}}{2\pi} d\omega_s = \frac{1}{2\pi\sqrt{-1}}\partial\bar\partial \log h_s \in C^\infty(\Lambda^{1,1}M) \, . \qquad (18.5.\text{d})$$

If (L_i, h_i) are holomorphic Hermitian line bundles, then $h = h_1 h_2$ is a Hermitian metric on $L_1 \otimes L_2$ and Equation (18.5.d) shows

$$c_1(L_1 \otimes L_2, h_1 \otimes h_2) = c_1(L_1, h_1) + c_1(L_2, h_2). \tag{18.5.e}$$

We defined $\bar{\partial}\{f_{I,J} s \otimes dz^I \otimes d\bar{z}^J\} = s \otimes \bar{\partial}(f_{I,J}) \wedge dz^I \wedge d\bar{z}^J$ in Equation (18.4.c); this is independent of the particular local holomorphic section which is chosen. However, were we to define ∂_L using a similar formula, it would not be well defined as the transition functions of L would enter. We adjust for this by adding a suitable 0th order term depending on the Hermitian fiber metric h to define ∂_h. Set

$$\partial_h = \text{ext}(dz^i)\nabla_{h,\partial_{z^i}} : C^\infty(L \otimes \Lambda^{p,q}) \to C^\infty(L \otimes \Lambda^{p,q+1}).$$

If s is a local non-vanishing holomorphic section to L, then

$$\partial_{h,s}(f_{I,J} s \otimes dx^I \wedge d\bar{z}^J) = s \otimes \{\partial f_{I,J} + f_{I,J}\partial\log(h_s)\} \wedge dz^I \wedge d\bar{z}^J.$$

We no longer have $\bar{\partial}\partial_h + \partial_h\bar{\partial} = 0$; instead the first Chern class of L enters. Assertion 1 in the following result is called the *Formula of Nakano* [51]. Assertion 3 is called the *Bochner–Kodaira–Nakano identity* [51].

Lemma 18.40   Let (L, h) be a holomorphic Hermitian line bundle over a Kähler manifold (M, g, J). Let $\Theta = \sqrt{-1}\bar{\partial}\partial\log h = 2\pi c_1(L, h)$.

1. $\partial_h\bar{\partial} + \bar{\partial}\partial_h = \text{ext}(\bar{\partial}\partial\log h)$.
2. $\bar{\partial}\,\text{int}(\Omega) - \text{int}(\Omega)\bar{\partial} = \sqrt{-1}\delta_h'$ and $\partial_h\text{int}(\Omega) - \text{int}(\Omega)\partial_h = -\sqrt{-1}\delta_h''$.
3. $\bar{\partial}\delta_h'' + \delta_h''\bar{\partial} = \partial_h\delta_h' + \delta_h'\partial_h + \{\text{ext}(\Theta)\,\text{int}(\Omega) - \text{int}(\Omega)\,\text{ext}(\Theta)\}$.

Proof. As noted above, $\bar{\partial}\partial\log h := \bar{\partial}\partial\log h_s$ is independent of s. We establish Assertion 1 by computing:

$$\partial_h\bar{\partial}(s \otimes f_{I,J}dz^I \wedge d\bar{z}^J) = \partial_h(s \otimes \bar{\partial}f_{I,J} \wedge dz^I \wedge d\bar{z}^J)$$
$$= s \otimes (\partial\bar{\partial}f_{I,J} + \partial\log h_s \wedge \bar{\partial}f_{I,J}) \wedge dz^I \wedge d\bar{z}^J,$$

$$\bar{\partial}\partial_h(s \otimes f_{I,J}dz^I \wedge d\bar{z}^J) = \bar{\partial}\{s \otimes (\partial f_{I,J} + f_{I,J}\partial\log h_s)dz^I \wedge d\bar{z}^J\}$$
$$= s \otimes \{\bar{\partial}\partial f_{I,J} + \bar{\partial}f_{I,J} \wedge \partial\log h_s + f_{I,J}\bar{\partial}\partial\log h_s\} \wedge dz^I \wedge dz^J$$
$$= s \otimes \{-\partial\bar{\partial}F_{I,J} - \partial\log h_s \wedge \bar{\partial}f_{I,J}\} \wedge dz^I \wedge d\bar{z}^J$$
$$+ \text{ext}(\bar{\partial}\partial\log h)\{s \otimes f_{I,J}dz^I \wedge d\bar{z}^J\}.$$

If L is the trivial line bundle with the flat metric, then Assertion 2 follows from Lemma 18.38. If we take a non-flat metric, the first derivatives of the fiber metric h on L enter. By Lemma 18.32, we may choose a holomorphic section for L where the first derivatives of the fiber metric h vanish. Consequently, these identities continue to hold true in the more general setting and Assertion 2 follows. This also shows that δ' is independent of h. We then apply the argument of Theorem 18.39 and compute:

$$\sqrt{-1}(\partial_h \delta_h' + \delta_h' \partial_h) = \partial_h \bar{\partial} \operatorname{int}(\Omega) - \partial_h \operatorname{int}(\Omega) \bar{\partial} + \bar{\partial} \operatorname{int}(\Omega) \partial_h - \operatorname{int}(\Omega) \bar{\partial} \partial_h,$$
$$-\sqrt{-1}(\bar{\partial} \delta_h'' + \delta_h'' \bar{\partial}) = \bar{\partial} \partial_h \operatorname{int}(\Omega) - \bar{\partial} \operatorname{int}(\Omega) \partial_h + \partial_h \operatorname{int}(\Omega) \bar{\partial} - \operatorname{int}(\Omega) \partial_h \bar{\partial}.$$

By Assertion 1, $\bar{\partial} \partial_h + \partial_h \bar{\partial} = \operatorname{ext}(\bar{\partial} \partial \log h)$, we may add these two identities to see

$$\sqrt{-1}(\partial_h \delta_h' + \delta_h' \partial_h) - \sqrt{-1}(\bar{\partial} \delta_h'' + \delta_h'' \bar{\partial}) = \operatorname{ext}(\bar{\partial} \partial \log h) \operatorname{int}(\Omega) - \operatorname{int}(\Omega) \operatorname{ext}(\bar{\partial} \partial \log h).$$

Assertion 3 now follows.   □

18.5.5 KODAIRA–AKIZUKI–NAKANO VANISHING THEOREM [1, 42].

If Ω is a real two-form with $J^* \Omega = \Omega$, set $g_\Omega(X, Y) := \Omega(X, JY)$. We have

$$g_\Omega(Y, X) = \Omega(Y, JX) = -\Omega(JY, X) = \Omega(X, JY) = g_\Omega(X, Y)$$

so g_Ω is a real symmetric two-tensor which is J-invariant; this correspondence identifies the bundle of symmetric J-invariant two-tensors with the bundle of anti-symmetric J-invariant two-tensors.

Let h be a Hermitian inner product on a holomorphic line bundle L over a complex manifold M. Let

$$c_1(L, h) = \frac{1}{2\pi \sqrt{-1}} \partial \bar{\partial} \log h$$

be the first Chern form and let $g_h(X, Y) := c_1(L, h)(X, JY)$ be the associated symmetric bilinear form:

$$g_h = \frac{1}{2\pi \sqrt{-1}} \partial_{z^i} \partial_{\bar{z}^j} \log(h) dz^i \otimes d\bar{z}^j.$$

We say that L is a *positive line bundle* if L admits an inner product so that g_h is positive definite; the associated Kähler form is then given by $c_1(L, h)$.

If M admits a positive line bundle, then necessarily M is Kähler. We will discuss these manifolds in more detail in Section 19.3.1. The following result will play a central role in that discussion and is the appropriate generalization of the Bochner Vanishing Theorem (Theorem 18.11) and the Lichnerowicz Vanishing Theorem (Theorem 18.23) to the holomorphic context. Let $\mathcal{O}_L$ be the sheaf of holomorphic sections to L. Theorem 18.31 yields $H^q(\mathcal{O}_L \otimes \Lambda^{p,0} M) = H^q_{\text{Dol}}(L \otimes \Lambda^{p,0} M) = \ker\{\Delta_{p,q,L}\}$.

Theorem 18.41 Let L be a positive line bundle over a holomorphic manifold M. Then $H^q(\mathcal{O}_{L \otimes \Lambda^{p,0} M}) = 0$ for $p + q > \mathfrak{m}$.

Proof. Let L be a positive line bundle. Then we can take the Kähler metric to be defined by L and set $\Theta = \Omega$ in Lemma 18.40. Let $u \in \ker\{\Delta_{p,q,L}\}$. By Lemma 18.40,

$$0 = (\bar{\partial}\delta''_h + \delta''_h\bar{\partial})u = (\partial_h\delta'_h + \delta'_h\partial_h)u + \{\text{ext}(\Theta)\,\text{int}(\Omega) - \text{int}(\Omega)\,\text{ext}(\Theta)\}u\,.$$

We take the L^2 inner product of this identity with u and integrate by parts to obtain:

$$\begin{aligned}
(\{\text{ext}(\Theta)\,&\text{int}(\Omega) - \text{int}(\Omega)\,\text{ext}(\Theta)\}\,u, u)_{L^2} \\
&= -((\partial_h\delta'_h + \delta'_h\partial_h)u, u)_{L^2} \\
&= -(\partial_h u, \partial_h u)_{L^2} - (\delta'_h u, \delta'_h u)_{L^2} \le 0\,.
\end{aligned} \tag{18.5.f}$$

On the other hand, if u is a form of degree (p, q), then Lemma 18.38 yields

$$\{\text{ext}(\Omega)\,\text{int}(\Omega) - \text{int}(\Omega)\,\text{ext}(\Omega)\}u = (p + q - \mathfrak{m})u\,.$$

If $p + q \ge \mathfrak{m}$, then we take the inner product of this identity with u to see

$$(\{\text{ext}(\Theta)\,\text{int}(\Omega) - \text{int}(\Omega)\,\text{ext}(\Theta)\}\,u, u)_{L^2} = (p + q - \mathfrak{m})(u, u)_{L^2} \ge 0\,. \tag{18.5.g}$$

We use Equations (18.5.f) and (18.5.g) to see $(u, u)_{L^2} = 0$ if $p + q > \mathfrak{m}$ and conclude that $u = 0$. $\qquad\square$

C H A P T E R   19

Complex Geometry

We shall discuss work of the following mathematicians, among others, in this chapter.

H. Cartan
(1904–2008)

S. Chern
(1911–2004)

P. Dolbeault
(1924–2015)

G. Fubini
(1879-1943)

E. Goursat
(1858–1936)

F. Hartogs
(1874–1943)

E. Kähler
(1906-2000)

P. Koebe
(1882–1945)

E. Kummer
(1810–1893)

R. Narasimhan
(1937–2015)

K. Oka
(1901–1978)

E. Pompeiu
(1873–1954)

E. Study
(1862–1930)

Section 19.1 is an introduction to multivariate holomorphic geometry; we shall assume the reader is familiar with the relevant material of Books I and II. Section 19.2 is an introduction to the geometry of complex projective space. In Section 19.3, we discuss Hodge manifolds. Section 19.4 treats the Kodaira Embedding Theorem; we show any compact holomorphic manifold admits a positive line bundle if and only if it embeds as a compact holomorphic submanifold of a complex projective space of some dimension. The elliptic operator theory of Chapter 17 and the analysis of Chapter 18 play a crucial role here.

19.1   INTRODUCTION TO HOLOMORPHIC GEOMETRY

In Section 19.1.1, we state the Uniformization Theorem; we shall omit the proof as there are many excellent modern treatments. In Section 19.1.2, we prove the Cauchy Integral Formula and discuss various consequences of it. In Section 19.1.3, we generalize the Cauchy Integral Formula by proving the Cauchy Integral Representation Formula. In Section 19.1.4, we solve the inhomogeneous Cauchy–Riemann equations $\partial_{\bar{z}} g = f$ on a manifold of the form $\mathcal{O} \times M$ where $\mathcal{O}$ is either an open ball in $\mathbb{C}$ or all of $\mathbb{C}$ and where M is an arbitrary smooth manifold; this will play an important role in our discussion of the Dolbeault Lemma subsequently in Section 19.1.6. In Section 19.1.5, we extend the Cauchy Integral Formula to the multivariate context. In Section 19.1.6, we prove the Dolbeault Lemma. Section 19.1.7 deals with the Theorem of Removable Singularities in the multi-variable context. Throughout this section, let $B_\varepsilon(z_0) := \{z \in \mathbb{C} : |z - z_0| < \varepsilon\}$ be the open ball of radius ε about the point z_0 in $\mathbb{C}$.

19.1.1 THE UNIFORMIZATION THEOREM. This result is a fundamental tool in the study of holomorphic surfaces. It was first proved by Koebe [44] and Poincaré [60] independently in 1907.

Theorem 19.1   If M is a simply connected Riemann surface, then M is holomorphically equivalent either to the open unit disk, to the complex plane, or to the two-dimensional sphere.

19.1.2 HOLOMORPHIC FUNCTIONS OF ONE VARIABLE. In the following lemma, Assertion 1 gives another proof of the smoothness result of Lemma 18.27, Assertion 2 is called the Cauchy Integral Formula, Assertion 3 deals with Taylor series, and Assertion 4 is called the Identity Theorem.

Lemma 19.2   Let $\mathcal{O}$ be a non-empty connected open subset of $\mathbb{C}$, let $f \in \mathrm{Hol}(\mathcal{O})$, and let $\bar{B}_\varepsilon(z_0) \subset \mathcal{O}$.

1. $f \in C^\infty(\mathcal{O})$.

2. If $z \in B_\varepsilon(z_0)$, then $f^{(n)}(z) = \dfrac{n!}{2\pi\sqrt{-1}} \displaystyle\int_{|\xi - z_0| = \varepsilon} \dfrac{f(\xi)}{(\xi - z)^{n+1}} d\xi$.

3. $\displaystyle\sum_{n=0}^{\infty} \dfrac{f^n(z_0)}{n!} (z - z_0)^n$ converges uniformly to f on compact subsets of $B_\varepsilon(z_0)$.

4. If f vanishes on any non-empty open subset of $\mathcal{O}$, then f vanishes on all of $\mathcal{O}$.

Proof. For $z \in B_\varepsilon(z_0)$ and r small, let $R := \{w : |w - z_0| \leq \varepsilon \cap \{w : |w - z| \geq r\}$ be the closed punctured disk pictured below:

By Lemma 18.27, $\frac{f(\xi)}{\xi - z} d\xi$ is a closed one-form. Thus, by Green's Theorem,

$$0 = \int_{\mathrm{bd}(R)} \frac{f(\xi)}{\xi - z} d\xi = \int_{|\xi - z_0| = \varepsilon} \frac{f(\xi)}{\xi - z} d\xi - \int_{|\xi - z| = r} \frac{f(\xi)}{\xi - z} d\xi. \qquad (19.1.\text{a})$$

We set $\xi = z + re^{\sqrt{-1}\theta}$. Then $\frac{d\xi}{\xi - z} = \sqrt{-1} d\theta$. Consequently,

$$\int_{|\xi - z| = r} \frac{f(\xi)}{\xi - z} d\xi = \sqrt{-1} \int_{\theta=0}^{2\pi} f(z + re^{\sqrt{-1}\theta}) d\theta.$$

As f is continuous, we can use Equation (19.1.a) and take the limit as $r \to 0$ to see

$$\frac{1}{2\pi\sqrt{-1}} \int_{|\xi - z_0| = \varepsilon} \frac{f(\xi)}{\xi - z} d\xi = \frac{1}{2\pi} \lim_{r \to 0} \int_{\theta=0}^{2\pi} f(z + re^{\sqrt{-1}\theta}) d\theta = f(z). \qquad (19.1.\text{b})$$

We differentiate Equation (19.1.b) to establish Assertions 1 and 2.

If $\xi \in \mathrm{bd}(B_\varepsilon(z_0))$ and $z \in B_\varepsilon(z_0)$, then $|(z - z_0)(\xi - z_0)^{-1}| < 1$. The following *geometric series* converges uniformly on compact subsets of $B_\varepsilon(z_0)$:

$$\frac{1}{\xi - z_0} \sum_{n=0}^{\infty} \frac{(z - z_0)^n}{(\xi - z_0)^n} = \frac{1}{\xi - z_0} \frac{1}{1 - \frac{z - z_0}{\xi - z_0}} = \frac{1}{\xi - z}. \qquad (19.1.\text{c})$$

We establish Assertion 3 by expanding

$$\begin{aligned}
f(z) &= \frac{1}{2\pi\sqrt{-1}} \int_{|\xi - z_0| = \varepsilon} \frac{f(\xi)}{\xi - z} d\xi = \frac{1}{2\pi\sqrt{-1}} \sum_{n=0}^{\infty} \int_{|\xi - z_0| = \varepsilon} \frac{(z - z_0)^n}{(\xi - z_0)^{n+1}} f(\xi) d\xi \\
&= \sum_{n=0}^{\infty} (z - z_0)^n \frac{1}{2\pi\sqrt{-1}} \int_{|\xi - z_0| = \varepsilon} \frac{1}{(\xi - z_0)^{n+1}} f(\xi) d\xi = \sum_{n=0}^{\infty} \frac{f^{(n)}(z_0)}{n!} (z - z_0)^n.
\end{aligned}$$

Suppose f vanishes identically in a neighborhood of some point $z_0 \in \mathcal{O}$. Then all the holomorphic derivatives of f vanish at z_0 so the Taylor series of f expanded around z_0 vanishes identically. By Assertion 3, this implies that f vanishes identically on the largest open disk about z_0 which is contained in $\mathcal{O}$. Since $\mathcal{O}$ is connected, we can find a finite sequence of open disks $\mathcal{O}_i$ joining z_0 to any other point z of $\mathcal{O}$ so that $\mathcal{O}_i \cap \mathcal{O}_{i+1}$ is non-empty. We use induction to see that if f vanishes on $\mathcal{O}_i$, then f vanishes on $\mathcal{O}_{i+1}$ and, consequently, $f(z) = 0$. Since z was arbitrary, Assertion 4 now follows. $\qquad \square$

Goursat [29] showed that the Cauchy Integral Formula holds under the weaker assumption that f is only assumed to be complex differentiable at each point of $\mathcal{O}$.

19.1.3 THE CAUCHY INTEGRAL REPRESENTATION FORMULA. The following result is called the *Cauchy Integral Representation Formula* or sometimes the *Cauchy–Pompeiu Formula* (see [61]) and can be regarded as a generalization of the Cauchy Integral Formula which was given Lemma 19.2. It can be used to solve the inhomogeneous Cauchy–Riemann equations. Let $\mathcal{O}$ be a bounded open subset of $\mathbb{C}$ with smooth boundary. If $f \in C^\infty(\overline{\mathcal{O}})$, define

$$g_f(z) := \frac{1}{2\pi\sqrt{-1}} \int_{\overline{\mathcal{O}}} \frac{f(\xi)}{\xi - z} d\xi \wedge d\bar{\xi} \quad \text{for} \quad z \in \mathcal{O}. \tag{19.1.d}$$

There is a technical point here. Although f is assumed smooth on the closure of $\mathcal{O}$, we only define g_f on $\mathcal{O}$. Since $\overline{\mathcal{O}}$ is compact and f is continuous on $\overline{\mathcal{O}}$, the only point at issue in verifying that Equation (19.1.d) is well defined arises from the apparent singularity at $\xi = z$ in the interior. If f is holomorphic, then Assertion 2 in the following lemma becomes the Cauchy Integral Formula.

Lemma 19.3   Assume that $f \in C^\infty(\overline{\mathcal{O}})$.

1. g_f is well defined and belongs to $C^\infty(\mathcal{O})$.
2. If $z \in \mathcal{O}$, then $f(z) = \dfrac{1}{2\pi\sqrt{-1}} \displaystyle\int_{\mathrm{bd}(\mathcal{O})} \frac{f(\xi)}{\xi - z} d\xi + \frac{1}{2\pi\sqrt{-1}} \int_{\overline{\mathcal{O}}} \frac{\partial_{\bar{\xi}} f(\xi)}{\xi - z} d\xi \wedge d\bar{\xi}.$
3. $\partial_{\bar{z}} g_f = f.$

Proof. We argue as follows to prove Assertion 1. Let $z_0 \in \mathcal{O}$. Choose $\delta > 0$ so that $B_{4\delta}(z_0) \subset \mathcal{O}$. Choose a plateau function $\vartheta \in C_0^\infty(B_{2\delta}(z_0))$ which is identically 1 on $B_\delta(z_0)$. We may decompose $g_f = g_{\vartheta f} + g_{(1-\vartheta)f}$. We use Equation (19.1.d) to see $g_{f(1-\vartheta)}$ is smooth on $B_\delta(z_0)$ since we have cut away the singularity. If $z \in B_\delta(z_0)$, introduce polar coordinates $\xi = z + re^{\sqrt{-1}\theta}$ centered at z. Since ϑ has support in $B_{2\delta}(z_0)$, we may assume $r \leq 3\delta$. Since

$$d\xi \wedge d\bar{\xi} = -2\sqrt{-1}dxdy = -2\sqrt{-1}rdrd\theta,$$

we have that

$$g_{\vartheta f}(z) = -\frac{1}{\pi} \int_{[0,3\delta]\times[0,2\pi]} f(z + re^{\sqrt{-1}\theta})\vartheta(z + re^{\sqrt{-1}\theta})e^{-\sqrt{-1}\theta}d\theta dr. \tag{19.1.e}$$

Consequently, $g_{\vartheta f} \in C^\infty(B_\delta(z_0))$ and hence $g \in C^\infty(B_\delta(z_0))$. Assertion 1 follows because the point z_0 of $\mathcal{O}$ was arbitrary. We apply Assertion 1 to see that

$$h(z) := \int_{\mathrm{bd}(\mathcal{O})} \frac{f(\xi)}{\xi - z} d\xi + \int_{\overline{\mathcal{O}}} \frac{\partial_{\bar{\xi}} f(\xi)}{\xi - z} d\xi \wedge d\bar{\xi}$$

is well defined and belongs to $C^\infty(\mathcal{O})$. By *Green's Theorem*,

$$-\int_{\xi\in\overline{\mathcal{O}},|\xi-z_0|\geq\delta} \frac{\partial_{\bar\xi} f}{\xi-z_0}d\xi\wedge d\bar\xi = \int_{\xi\in\overline{\mathcal{O}},|\xi-z_0|\geq\delta} d\left\{\frac{f(\xi)}{\xi-z_0}d\xi\right\}$$

$$= \int_{\mathrm{bd}(\mathcal{O})} \frac{f(\xi)}{\xi-z_0}d\xi - \int_{|\xi-z_0|=\delta} \frac{f(\xi)}{\xi-z_0}d\xi.$$

(19.1.f)

We set $\xi = z_0 + \delta e^{\sqrt{-1}\theta}$ and use Equation (19.1.f) to prove Assertion 2 by verifying

$$
\begin{aligned}
h(z_0) &= \int_{\mathrm{bd}(\mathcal{O})} \frac{f(\xi)}{\xi-z_0}d\xi + \lim_{\delta\to 0}\int_{\xi\in\overline{\mathcal{O}},|\xi-z_0|\geq\delta} \frac{\partial_{\bar\xi} f(\xi)}{\xi-z_0}d\xi\wedge d\bar\xi \\
&= \lim_{\delta\to 0}\int_{|\xi-z_0|=\delta} \frac{f(\xi)}{\xi-z_0}d\xi = \lim_{\delta\to 0}\int_{\theta=0}^{2\pi} \frac{f(z_0+\delta e^{\sqrt{-1}\theta})}{\delta e^{\sqrt{-1}\theta}}\delta\sqrt{-1}e^{\sqrt{-1}\theta}d\theta \\
&= \lim_{\delta\to 0}\int_{\theta=0}^{2\pi} f(z_0+\delta e^{\sqrt{-1}\theta})\sqrt{-1}d\theta = 2\pi\sqrt{-1}f(z_0).
\end{aligned}
$$

We establish Assertion 3 as follows. Let $z_0\in\mathcal{O}$. Expand

$$g_f = g_{\vartheta f} + g_{(1-\vartheta)f}.$$

Since $(1-\vartheta)$ vanishes on $B_\delta(z_0)$, we may differentiate Equation (19.1.d) under the integral sign to see $\partial_{\bar z}\{g_{(1-\vartheta)f}\} = 0$ on $B_\delta(z_0)$. Thus, $\partial_{\bar z}\{g_f\} = \partial_{\bar z}\{g_{\vartheta f}\}$ on $B_\delta(z_0)$. We use Equation (19.1.e) to differentiate under the integral sign and verify $\partial_{\bar z}\{g_{\vartheta f}\} = g_{\partial_{\bar z}\{\vartheta f\}}$ on $B_\delta(z_0)$. Since ϑ vanishes on the boundary of $\mathcal{O}$, we apply Assertion 2 to see if $z\in B_\delta(z_0)$, then

$$
\begin{aligned}
\partial_{\bar z}\{g_f\}(z) &= \partial_{\bar z}\{g_{\vartheta f}\}(z) = \frac{1}{2\pi\sqrt{-1}}\int_{\overline{\mathcal{O}}} \frac{\{\partial_{\bar\xi}(\vartheta f)\}(\xi)}{\xi-z}d\xi\wedge d\bar\xi \\
&= \frac{1}{2\pi\sqrt{-1}}\left\{\int_{\mathrm{bd}(\mathcal{O})} \frac{f(\xi)\vartheta(\xi)}{\xi-z}d\xi + \int_{\overline{\mathcal{O}}} \frac{\partial_{\bar\xi}\{\vartheta f\}(\xi)}{\xi-z}d\xi\wedge d\bar\xi\right\} \\
&= (\vartheta f)(z) = f(z).
\end{aligned}
$$

Since z_0 was arbitrary, Assertion 3 holds. □

19.1.4 THE EQUATION $\partial_{\bar z}g = f$. We extend Lemma 19.3 to the multi-variable context. We also remove the assumption that f extends smoothly to the boundary.

Lemma 19.4   Let $\mathcal{U} = B_r(z_0)$ or $\mathcal{U} = \mathbb{C}$, let $\mathcal{V}$ be an auxiliary open subset of $\mathbb{C}^{m-1}$, and let $f(z,\vec{w})\in C^\infty(\mathcal{U}\times\mathcal{V})$. There exists $g\in C^\infty(\mathcal{U}\times\mathcal{V})$ so that $\partial_{\bar z}g = f$ and so that if f is holomorphic in w^i, then g is holomorphic in w^i.

Proof. If $\mathcal{U} = B_r(z_0)$, let $\mathcal{U}_n := B_{r(1-\frac{1}{n})}(z_0)$. If $\mathcal{U} = \mathbb{C}$, let $\mathcal{U}_n = B_n(0)$; the sets $\overline{\mathcal{U}}_n$ form a compact exhaustion of $\mathcal{U}$. Let K_n be a compact exhaustion of $\mathcal{V}$. Let $\vartheta_n \in C_0^\infty(\mathcal{U}_n)$ be identically 1 on $\mathcal{U}_{n-1}$. Since $\vartheta_n f$ vanishes on the boundary of $\mathcal{U}_n$, there is no need to add a boundary correction term. We generalize Equation (19.1.d) and define

$$g_n(z, w) := \frac{1}{2\pi\sqrt{-1}} \int_{\mathcal{U}_n} \frac{\vartheta_n(\xi) f(\xi, w)}{\xi - z} d\xi \wedge d\bar{\xi} \quad \text{for} \quad (z, w) \in \mathcal{U}_n \times K_n. \qquad (19.1.g)$$

The argument given to prove Lemma 19.3 then extends to show

$$\partial_{\bar{z}} g_n = \vartheta_n f = f \quad \text{on} \quad \mathcal{U}_{n-1} \times K_n.$$

Let $n \geq 3$. We have $\partial_{\bar{z}}(g_n - g_{n+1}) = 0$ on $\mathcal{U}_{n-1} \times K_n$ so $g_n - g_{n+1}$ is holomorphic for z belonging to $\mathcal{U}_{n-1}$. Let

$$a_{n,k}(w) := \frac{1}{2\pi\sqrt{-1}} \int_{\mathrm{bd}(\mathcal{U}_{n-1})} \frac{g_{n+1}(\xi, w) - g_n(\xi, w)}{\xi^{k+1}} d\xi. \qquad (19.1.h)$$

The argument given to prove Lemma 19.2 then shows the power series

$$\sum_{k=0}^{\infty} a_{n,k}(w) z^k \qquad (19.1.i)$$

converges uniformly to $g_n - g_{n+1}$ on $\mathcal{U}_{n-2} \times K_n$ and, furthermore, that all the w derivatives also converge uniformly. Let $\|\cdot\|_n$ be the C^n norm on $C^\infty(\mathcal{U}_n \times K_n)$. By approximating the difference $g_n - g_{n+1}$ by a partial sum of the series in Equation (19.1.i), we can choose $h_n(z, w)$ to be smooth in w, polynomial in z, and satisfy $\|g_n - g_{n+1} - h_n\|_{n-2} < 2^{-n}$. Thus, by replacing g_{n+1} by $g_{n+1} + h_n$, we may assume the sequence g_n is chosen so $\partial_{\bar{z}} g_{n+1} = f$ on $\mathcal{U}_{n-2} \times K_n$ and $\|g_n - g_{n+1}\|_{n-2} < 2^{-n}$. Let $g_\infty = \lim_{n\to\infty} g_n \in C^\infty(\mathcal{U} \times \mathcal{V})$. We then have that $\partial_{\bar{z}} g_\infty = f$ on $\mathcal{U} \times \mathcal{V}$ as desired. If f is holomorphic in w^i, we use Equation (19.1.g) to see g_n is holomorphic in w^i. We use Equation (19.1.h) to see $a_{n,k}$ is holomorphic in w^i and hence the partial Taylor series h_n will be holomorphic in w^i. Since the uniform limit of holomorphic functions is holomorphic, g_∞ is holomorphic in w^i. $\qquad \square$

Remark 19.5   We elected to use Lemma 19.3 to prove Lemma 19.4. One can, however, use Fourier Series. We use the estimates of Example 16.33 or, equivalently, the results of Section 17.4.4, to justify the following formal calculation. Let $\mathbb{T}^2 := \mathbb{R}^2/(2\pi\mathbb{Z})^2$ be the square torus; embed $\mathcal{O}_{n+1}$ holomorphically as an open subset of $\mathbb{T}^2$. Expand

$$\vartheta_n(z) f(z, w) = \sum_{n,m} a_{n,m}(w) e^{\sqrt{-1}(mx+ny)}$$

where

$$a_{m,n}(w) := \frac{1}{2\pi} \int_{\mathbb{T}^2} \vartheta(z) f(z) e^{-\sqrt{-1}(mx+ny)} dx dy.$$

Define $g_n(z, w) \in C^\infty(\mathcal{O}_{n+1} \times K_n)$ by setting:

$$g_n(z, w) := a_{0,0}(w)\bar{z} + \sum_{(m,n) \neq (0,0)} \frac{2a_{m,n}(z, w)}{\sqrt{-1}n + m} e^{\sqrt{-1}(mx+ny)}.$$

One then has $\partial_{\bar{z}}\{g_n\} = \vartheta_n f$ on $\mathcal{O}_{n+1} \times K_n$. The remainder of the argument is the same.

19.1.5 HOLOMORPHIC FUNCTIONS OF SEVERAL VARIABLES. If $\varepsilon_i > 0$, let

$$B_{\vec{\varepsilon}}(\vec{z}_0) := B_{\varepsilon_1}(z_0^1) \times \cdots \times B_{\varepsilon_m}(z_0^m)$$

be the open *poly-disk*. Let $S_{\vec{\varepsilon}}(\vec{z}_0) = \mathrm{bd}(B_{\varepsilon_1}(z_0^1)) \times \cdots \times \mathrm{bd}(B_{\varepsilon_m}(z_0^m))$. The following result is an immediate consequence of Lemmas 18.27 and 19.2.

Theorem 19.6   Let $\mathcal{O}$ be a non-empty connected open subset of $\mathbb{C}^m$. Let $f \in \mathrm{Hol}(\mathcal{O})$. If z_0 belongs to $\mathcal{O}$, choose $\vec{\varepsilon}$ so that $\bar{B}_{\vec{\varepsilon}}(\vec{z}_0) \subset \mathcal{O}$.

1. If $\vec{z} \in B_{\vec{\varepsilon}}(\vec{z}_0)$, then $f(\vec{z}) = \dfrac{1}{(2\pi\sqrt{-1})^m} \displaystyle\int_{S_{\vec{\varepsilon}}(\vec{z}_0)} \dfrac{f(\vec{\xi})}{(\xi^1 - z^1) \cdot \cdots \cdot (\xi^m - z^m)} d\xi^1 \cdots d\xi^m$.

2. f is smooth and has holomorphic derivatives of all orders on $\mathcal{O}$.

3. The multivariate Taylor series

$$\sum_{i_1=0,\ldots,i_m=0}^{\infty} \frac{f^{(i_1,\ldots,i_m)}(\vec{z}_0)}{i_1! \cdot \cdots \cdot i_m!} (z^1 - z_0^1)^{i_1} \cdot \cdots \cdot (z^m - z_0^m)^{i_m}$$

converges uniformly and absolutely to f on compact subsets of $B_{\vec{\varepsilon}}(\vec{z}_0)$.

4. If f vanishes on any non-empty open subset of $\mathcal{O}$, then f vanishes identically.

19.1.6 THE POINCARÉ [59] AND DOLBEAULT [20] LEMMAS. Let

$$I_{\vec{\varepsilon}}(\vec{x}_0) := \{\vec{x} \in \mathbb{R}^m : |x^i - x_0^i| < \varepsilon^i \text{ for } 1 \leq i \leq m\},$$
$$B_{\vec{\varepsilon}}(\vec{z}_0) := \{\vec{z} \in \mathbb{C}^m : |z^i - z_0^i| < \varepsilon^i \text{ for } 1 \leq i \leq m\},$$

be an open rectangle in $\mathbb{R}^m$ and an open *poly-disk* in $\mathbb{C}^m$, respectively.

Theorem 19.7

1. **(Poincaré Lemma)**. Let $\omega \in C^\infty(\Lambda^q I_{\vec{\varepsilon}}(\vec{x}_0))$ for $q > 0$. If $d\omega = 0$, then there exists ϕ in $C^\infty(\Lambda^{q-1} I_{\vec{\varepsilon}}(\vec{x}_0))$ such that $d\phi = \omega$.

2. **(Dolbeault Lemma)**. Let $\omega \in C^\infty(\Lambda^{p,q} B_{\vec{\varepsilon}}(\vec{x}_0))$ for $q > 0$. If $\bar{\partial}\omega = 0$, then there exists ϕ in $C^\infty(\Lambda^{p,q-1} B_{\vec{\varepsilon}}(\vec{x}_0))$ such that $\bar{\partial}\phi = \omega$.

Proof. Let $I = \{1 \leq i_1 < \cdots < i_q \leq m - 1\}$ and $J = \{1 \leq j_1 < \cdots < j_{q-1} < m - 1\}$ be collections of indices which are at most $m - 1$. Decompose

$$\omega = \sum_{|I|=q} f_I dx^I + \sum_{|J|=q-1} g_J dx^m \wedge dx^J .$$

Solve the equation $\partial_{x^m} h_J = g_J$ by setting

$$h_J(x^1, \ldots, x^m) := \int_{t=x_0^m - \varepsilon_m}^{x^m} g_J(x^1, \ldots, x^{m-1}, t) dt . \tag{19.1.j}$$

By replacing ϕ by $\phi - d(h_J dx^J)$, we may assume that ϕ does not involve dx^m. Because $d\phi$ vanishes, ϕ does not depend on x^m either and we have eliminated the variable x^m. We continue to eliminate variables successively; we eventually must have $\phi = 0$ since $q > 0$. This completes the proof of Assertion 1. We argue similarly to prove Assertion 2 using Lemma 19.4 instead of Equation (19.1.j). The holomorphic indices play no role so we may set $p = 0$. Let

$$I = \{1 \leq i_1 < \cdots < i_q \leq \mathfrak{m} - 1\} \quad \text{and} \quad J = \{1 \leq j_1 < \cdots < j_{q-1} \leq \mathfrak{m} - 1\}$$

be collections of indices which are at most $\mathfrak{m} - 1$. Decompose

$$\omega = \sum_{|I|=q} f_I d\bar{z}^I + \sum_{|J|=q-1} g_J d\bar{z}^{\mathfrak{m}} \wedge d\bar{z}^J .$$

We apply Lemma 19.4 to choose h_J so that $\partial_{\bar{z}^{\mathfrak{m}}} h_J = g_J$. Replace ω by $\omega - \bar{\partial}(h_J d\bar{z}^J)$ to assume ω does not involve $d\bar{z}^{\mathfrak{m}}$. Since $\bar{\partial}\omega = 0$, the f_I are "independent of $\bar{z}^{\mathfrak{m}}$," i.e., the f_I are holomorphic in $z^{\mathfrak{m}}$. This condition is preserved when we apply Lemma 19.4 again to eliminate $d\bar{z}^{\mathfrak{m}-1}$. We continue in this fashion to eliminate variables and obtain $\omega = 0$. □

Theorem 19.7 is central to the proof that the Dolbeault cohomology groups $H^{p,q}(M)$ are isomorphic to the sheaf cohomology groups in the sheaf Ω^p of holomorphic differential forms of type $(p, 0)$, i.e., $H^{p,q}(M) \approx H^q(M, \Omega^p)$. We shall omit details as this is tangential to our line of inquiry. It permits one to complete the proof of Theorem 18.31 by identifying the Dolbeault cohomology groups defined analytically with the corresponding sheaf-theoretic groups exactly as was done in identifying the de Rham cohomology groups with sheaf-theoretic groups in Section 8.4 of Book II in the real setting; we refer to Dolbeault [20] and Serre [67] for further details.

19.1.7 REMOVABLE SINGULARITIES. The following results fall under general rubric of *Hartogs' Lemma* [35, 36]. Assertion 2 of the following result is called the *Theorem of Removable Singularity*. It shows that if $N^{\mathfrak{m}-1}$ is a holomorphic submanifold of a holomorphic manifold $M^{\mathfrak{m}}$ of complex codimension 1 and if f is holomorphic and bounded on $M^{\mathfrak{m}} - N^{\mathfrak{m}-1}$, then f extends holomorphically to all of $M^{\mathfrak{m}}$. It will play an important role in our proof of Lemma 19.20 which shows that the blowup is unique and in motivating the use of the blowup in the proof of the Kodaira Vanishing Theorem in Section 19.4.

Lemma 19.8   Let $\mathfrak{m} > 1$ and let $s > 4\mathfrak{m}$. If $s = \infty$, set $B_s(0) = \mathbb{C}^\mathfrak{m}$.

1. Let K be a compact non-empty subset of $B_1(0)$ with $B_1(0) - K$ connected. Then the restriction map $r : \mathrm{Hol}(B_s(0)) \to \mathrm{Hol}(B_s(0) - K)$ is an isomorphism.

2. Let $\mathcal{O}$ be an open connected subset of $\mathbb{C}^{\mathfrak{m}-1}$. Let $f \in \mathrm{Hol}((B_1(0) - \{0\}) \times \mathcal{O})$. If f is bounded, then there exists $\tilde{f} \in \mathrm{Hol}(B_1(0) \times \mathcal{O})$ so $r(\tilde{f}) = f$.

Proof. Let $\mathcal{U}_r$ be the poly-disk $B_r(0) \times \cdots \times B_r(0) \subset \mathbb{C}^\mathfrak{m}$. By the triangle inequality,

$$B_1(0) \subset \mathcal{U}_1 \subset \mathcal{U}_4 \subset B_{4\mathfrak{m}}(0) \subset B_s(0).$$

Let $f \in \mathrm{Hol}(B_s(0) - K)$. Define $F \in \mathrm{Hol}(\mathcal{U}_2)$ by

$$F(\vec{z}) := (2\pi\sqrt{-1})^{-\mathfrak{m}} \int_{|\xi^1|=2r} \cdots \int_{|\xi^\mathfrak{m}|=2r} \frac{f(\xi^1,\dots,\xi^\mathfrak{m})}{(\xi^1-z^1)\cdot\,\cdots\,\cdot(\xi^\mathfrak{m}-z^\mathfrak{m})} d\xi^\mathfrak{m} \cdots d\xi^1.$$

Since $|\xi^i| = 2r$ for $1 \le i \le \mathfrak{m}-1$, $(\xi^1,\dots,\xi^{\mathfrak{m}-1},z^\mathfrak{m}) \in B_s(0) - K$ if $1 \le |z^\mathfrak{m}| \le 4$. Consequently, Cauchy Integral Formula (see Lemma 19.2) yields

$$f(\xi^1,\dots,\xi^{\mathfrak{m}-1},z^\mathfrak{m}) = (2\pi\sqrt{-1})^{-1} \int_{|\xi^\mathfrak{m}|=2r} \frac{f(\xi^1,\dots,\xi^{\mathfrak{m}-1},\xi^\mathfrak{m})}{\xi^\mathfrak{m}-z^\mathfrak{m}} d\xi^\mathfrak{m}.$$

Let $\mathcal{A} := \{\vec{z} : |z^1| < r,\dots,|z^{\mathfrak{m}-1}| < r, r < |z^\mathfrak{m}| < 2r\} \subset B_s(0) - K$. By Theorem 19.6,

$$
\begin{aligned}
F(\vec{z}) &= (2\pi\sqrt{-1})^{\mathfrak{m}-1} \int_{|\xi^1|=2r} \cdots \int_{|\xi^{\mathfrak{m}-1}|=2r} \frac{f(\xi^1,\dots,\xi^{\mathfrak{m}-1},z^\mathfrak{m})}{(\xi^1-z^1)\cdot\,\cdots\,\cdot(\xi^{\mathfrak{m}-1}-z^{\mathfrak{m}-1})} d\xi^{\mathfrak{m}-1} \cdots d\xi^1 \\
&= f(\vec{z}) \quad \text{if} \quad \vec{z} \in \mathcal{A}.
\end{aligned}
$$

The assumption $B_1(0) - K$ is connected implies $B_s(0) - K$ is connected. Since $\mathcal{A}$ is a non-empty open subset of $B_s(0) - K$ and since $B_s(0) - K$ is connected, the Identity Theorem (see Theorem 19.6) shows that $F = f$ on $B_s(0) - K$ which proves Assertion 1.

We extend the usual proof from the one-variable case to the multi-variable setting to establish Assertion 2. Suppose $f(z, \vec{w})$ is holomorphic and bounded for $0 < |z| < 4r$ and $\vec{w}$ in $\mathcal{O} \subset \mathbb{C}^{\mathfrak{m}-1}$. We set

$$F(z, \vec{w}) := \frac{1}{2\pi\sqrt{-1}} \int_{|\xi|=3r} \frac{f(\xi, \vec{w})}{\xi - z} d\xi \in \mathrm{Hol}(B_{3r}(0) \times \mathcal{O}).$$

Suppose that $\varepsilon < 2r < |z| < 3r$. Since $\frac{f(\xi, \vec{w})}{\xi-z}$ is holomorphic on the closed doubly punctured disk $\{\xi : |\xi| \le 3r, |\xi - z| \ge \varepsilon, |\xi| \ge \varepsilon\}$, Green's Theorem lets us deform the contour $|\xi| = 3r$ to the two contours $|\xi| = \varepsilon$ and $|\xi - z| = \varepsilon$ to express $F = F_1 + F_2$ where

$$F_1(\xi, \vec{w}) := \frac{1}{2\pi\sqrt{-1}} \int_{|\xi|=\varepsilon} \frac{f(\xi, \vec{w})}{\xi - z} d\xi \quad \text{and} \quad F_2(\xi, \vec{w}) := \frac{1}{2\pi\sqrt{-1}} \int_{|\xi-z|=\varepsilon} \frac{f(\xi, \vec{w})}{\xi - z} d\xi.$$

The contours in question are pictured below:

We use the Cauchy Integral Formula to see $F_2 = f(z, \vec{w})$. By assumption, $|f(z, \vec{w})| \leq C$ for some constant C. Since $|z| > 2r$, we have $|f(\xi, \vec{w})(\xi - z)^{-1}| \leq C_1$ if $|\xi| = \varepsilon < r$. We may set $\xi = \varepsilon e^{\sqrt{-1}\theta}$. Then $d\xi = \varepsilon \sqrt{-1} e^{\sqrt{-1}\theta} d\theta$ so

$$\left| \frac{1}{2\pi\sqrt{-1}} \int_{|\xi|=\varepsilon} \frac{f(\xi, \vec{w})}{\xi - z} d\xi \right| \leq \frac{1}{2\pi} \int_{\theta=0}^{2\pi} C\varepsilon\, d\theta \quad \text{so} \quad \lim_{\varepsilon \to 0} \frac{1}{2\pi\sqrt{-1}} \int_{|\xi|=\varepsilon} \frac{f(\xi, \vec{w})}{\xi - z} d\xi = 0.$$

It now follows $F(z, \vec{w}) = f(z, \vec{w})$ for $r < |z| < 2r$ and $\vec{w} \in \mathcal{O}$. Since this is a non-empty subset of $B_1(0) \times \mathcal{O}$ and since $B_1(0) \times \mathcal{O}$ is connected, the Identity Theorem (see Theorem 19.6) shows that $F(z, \vec{w}) = f(z, \vec{w})$ on $\{\mathcal{O}_1 - \{0\}\} \times \mathcal{O}_2$. Consequently, F provides the desired extension. □

19.2   THE GEOMETRY OF COMPLEX PROJECTIVE SPACE

In Section 19.2.1, we define complex projective space. We define the tautological line bundle $\mathbb{L} = \mathbb{L}_{\mathbb{C}}$ over complex projective space in Section 19.2.2. We use the first Chern form of $\mathbb{L}$ to discuss various geometrical properties of the Fubini–Study metric in Section 19.2.3.

19.2.1   COMPLEX PROJECTIVE SPACE.

Complex projective space $\mathbb{CP}^m$ is a particular example of a holomorphic manifold which will play a central role in our discussion. It is the space of complex lines through the origin in $\mathbb{C}^{m+1}$ (see Section 4.3.3 of Book II). If $\vec{z} = (z^1, \dots, z^{m+1}) \in \mathbb{C}^{m+1} - \{0\}$, let $\langle \vec{z} \rangle := \vec{z} \cdot \mathbb{C} \in \mathbb{CP}^m$ be the corresponding complex line determined by $\vec{z}$. We have $\langle \vec{z} \rangle = \langle \vec{w} \rangle$ if and only if $\vec{z}$ is a non-zero complex multiple of $\vec{w}$. Thus, we may identify $\mathbb{CP}^m$ with the quotient $\{\mathbb{C}^{m+1} - \{0\}\}/\{\mathbb{C} - \{0\}\}$ to introduce homogeneous coordinates $\langle z^1, \dots, z^{m+1} \rangle$ on $\mathbb{CP}^m$. Restricting to unit vectors $\vec{w}$, we may also identify $\mathbb{CP}^m$ with the quotient S^{2m+1}/S^1; this gives rise to the *Hopf fibration*

$$S^1 \to S^{2m+1} \overset{\pi}{\longrightarrow} \mathbb{CP}^m$$

and shows, in particular, that $\mathbb{CP}^m$ is compact where we give $\mathbb{CP}^m$ the *quotient topology* (see Section 16.1.1). The unitary group $U(m + 1)$ acts on $\mathbb{CP}^m$ by

$$\Phi \cdot \langle \vec{z} \rangle = \langle \Phi\vec{z} \rangle \quad \text{for} \quad \Phi \in U(m + 1).$$

As the corresponding action on S^{2m+1} is continuous, this is a continuous action on $\mathbb{CP}^m$. The isotropy subgroup of this action is given by $U(m) \times U(1)$ and, consequently, we also have a fibration

$$U(m) \times U(1) \to U(m + 1) \to \mathbb{CP}^m$$

which identifies $\mathbb{CP}^m$ with the coset space $U(m+1)/\{U(m) \times U(1)\}$. This shows that $\mathbb{CP}^m$ is a *homogeneous space* and the analysis of Section 7.1 of Book II pertains. Since $U(m+1)$ is compact, we can average over the unitary group to define $U(m+1)$-invariant metrics. Since $U(m+1)$ acts transitively on the projective tangent bundle $(T(\mathbb{CP}^m) - 0)/\{\mathbb{R} - \{0\}\}$, a $U(m+1)$-invariant Riemannian metric on $\mathbb{CP}^m$ is determined up to scale; in Section 19.2.3, we will discuss the *Fubini–Study metric* which was defined by Fubini [24] and Study [71]; this Riemannian metric on $\mathbb{CP}^m$ is $U(m+1)$-invariant.

Give $\mathbb{CP}^m$ the Fubini–Study metric. Since $U(m+1)$ acts transitively on the unit tangent bundle of $\mathbb{CP}^m$, $\mathbb{CP}^m$ is a *two-point homogeneous space*; given 4 points P_i in $\mathbb{CP}^m$ with the geodesic distance from P_1 to P_2 equal to the geodesic distance from P_3 to P_4, there exists $\Phi \in U(m+1)$ defining an isometry of $\mathbb{CP}^m$ so that $\Phi P_1 = P_3$ and $\Phi P_2 = P_4$.

We can discuss the holomorphic structure on $\mathbb{CP}^m$ very concretely. We define an open cover of $\mathbb{CP}^m$ by setting $\mathcal{O}_j := \{\langle \vec{z} \rangle : z^i \neq 0\}$. We put a "1" in the jth position to define the map Ψ_j sending $(w^1, \ldots, w^m)$ to $\langle w^1, \ldots, 1, \ldots, w^m \rangle$ to identify $\mathcal{O}_j$ with $\mathbb{C}^m$ and introduce holomorphic coordinates on $\mathcal{O}_j$. We set $z_j^i := \frac{z^i}{z^j}$ and $\Psi_j^{-1}\{\langle z^1, \ldots, z^{m+1} \rangle\} := (z_j^1, \ldots, z_j^m)$ where we omit $z_i^i = 1$. Since $z_i^i = \frac{z_k^i}{z_k^j}$, the transition functions are holomorphic and $\mathbb{CP}^m$ is a holomorphic manifold. Take the standard basis $\{e_1, \ldots, e_{m+1}\}$ for $\mathbb{C}^{m+1}$. Let Φ belong to $GL(m+1, \mathbb{C})$. We may expand $\Phi e_i = \Phi_i^j e_j$. We then have

$$\Phi(\vec{z}) = \Phi(z^i e_i) = \Phi_i^j z^i e_j = (\Phi_i^1 z^i, \ldots, \Phi_i^{m+1} z^i).$$

One then verifies that the induced map on $\mathbb{CP}^m$ is holomorphic. This is best illustrated by taking $m = 1$. Let $\Phi(z^1, z^2) = (\alpha z^1 + \beta z^2, \gamma z^1 + \delta z^2)$. The induced map on $\mathcal{O}_1$ is the *linear fractional transformation*

$$w \to \langle 1, w \rangle \to \langle \alpha + \beta w, \gamma + \delta w \rangle \to \frac{\gamma + \delta w}{\alpha + \beta w}.$$

It is singular at the point $\alpha + \beta w = 0$; this apparent singularity can be eliminated by examining the induced map from $\mathcal{O}_2$ to $\mathcal{O}_2$ which is defined by

$$w \to \frac{\alpha + \beta w}{\gamma + \delta w}.$$

19.2.2 THE TAUTOLOGICAL LINE BUNDLE.

There is a natural line bundle $\mathbb{L}$ over $\mathbb{CP}^m$ which is called the *classifying line bundle* or the *tautological line bundle*. The fiber of $\mathbb{L}$ over a point P of $\mathbb{CP}^m$ is the complex line which P represents. More precisely, $\mathbb{L}$ is defined by setting:

$$\mathbb{L} := \{(\langle \vec{z} \rangle, \vec{w}) \in \mathbb{CP}^m \times \mathbb{C}^{m+1} : \vec{w} \in \langle \vec{z} \rangle\}.$$

Over $\mathcal{O}_j$, we define a local section setting

$$s_j(\langle \vec{z} \rangle) = \left(\langle \vec{z} \rangle \times \left(\frac{z^1}{z^j}, \ldots, \frac{z^{m+1}}{z^j} \right) \right). \tag{19.2.a}$$

We then have

$$s_j = z_j^i s_i \quad \text{on} \quad \mathcal{O}_i \cap \mathcal{O}_j \tag{19.2.b}$$

so the transition functions are holomorphic and $\mathbb{L}$ is a holomorphic line bundle; the sections s_j can be regarded as meromorphic sections which blowup along the hyperplanes $z^j = 0$. Let $1 := \mathbb{CP}^m \times \mathbb{C}$ be the trivial line bundle. The following result played an important role in our computation of the Stiefel–Whitney classes of $\mathbb{CP}^m$ in Example 18.20. Let $T_c(\mathbb{CP}^m)$ be the complex tangent bundle.

Lemma 19.9   There exist holomorphic short exact sequences:

1. $0 \to \Lambda^{1,0}(\mathbb{CP}^m) \to \mathbb{L} \otimes 1^{m+1} \to 1 \to 0$.

2. $0 \leftarrow T_c(\mathbb{CP}^m) \leftarrow \mathbb{L}^* \otimes 1^{m+1} \leftarrow 1 \leftarrow 0$.

3. $\Lambda^{m,0}\mathbb{CP}^m$ is holomorphically isomorphic to $\mathbb{L}^{m+1}$.

4. Let $\pi : \mathbb{L} \to \mathbb{CP}^m$ be the canonical projection. Then $\Lambda^{m+1,0}(\mathbb{L})$ is holomorphically isomorphic to $\pi^*(\mathbb{L})^m$.

Proof. We do not sum over repeated indices in what follows. We summarize the notation and argue as follows to prove Assertion 1. As the argument is somewhat combinatorial, we give it in outline form for the sake of clarity. We use Equation (19.2.b).

1. Local holomorphic coordinates $z_k^j = \frac{z^j}{z^k}$.

2. Local holomorphic sections $s_k((\vec{z})) = ((\vec{z}) \times (z_k^1, \ldots, z_k^{m+1}))$ to $\mathbb{L}$ over $\mathcal{O}_k$.

3. Local frame $\vec{s}_j = (s_j^1, \ldots, s_j^{m+1})$ for $\mathbb{L} \otimes 1^{m+1}$ over $\mathcal{O}_j$.

4. On $\mathcal{O}_j \cap \mathcal{O}_k$, $z_j^k \cdot z_k^\ell = z_j^\ell$, $dz_j^i = (z_k^j)^{-1} dz_k^i - (z_k^j)^{-2} z_k^i dz_k^j$ and $z_j^k \cdot s_k^i = s_j^i$.

5. Set $F_j(dz_j^i) := s_j^i - z_j^i s_j^j$ mapping $\Lambda^{1,0}\mathcal{O}_j$ to $\mathbb{L} \otimes 1^{n+1}$. Then F_j is well defined since $dz_j^j = s_j^j - z_j^j s_j^j = 0$.

6. On $\mathcal{O}_j \cap \mathcal{O}_k$, we use (4) to compute $F_k(dz_j^i) = (z_k^j)^{-1} F_k(dz_k^i) - (z_k^j)^{-2} z_k^i F_k(dz_k^j)$
    $$= (z_k^j)^{-1} s_k^i - (z_k^j)^{-1} z_k^i s_k^k - (z_k^j)^{-2} z_k^i s_k^j + (z_k^j)^{-2} z_k^i z_k^j s_k^k = s_j^i - z_j^i s_j^j = F_j(dz_j^i).$$

7. By (6), the collection $\{F_j\}$ defines a global holomorphic map
    $$F : \Lambda^{1,0}M \to L \otimes 1^{m+1}.$$

8. Since only $F_j(dz_j^i)$ involves s_j^i, F is injective.

We introduce some additional notation. Let $\text{coker}\{F\} := L \otimes 1^{m+1} / \text{range}\{F\}$ be the quotient bundle. Let $\pi : L \otimes 1^{m+1} \to \text{coker}\{F\}$ be the natural projection.

9. Since s_j^j does not belong to $\text{range}\{F_j\}$, $\pi(s_j^j) \neq 0$.

10. $F(dz_j^k) = s_j^k - z_j^k s_j^j = z_j^k(s_k^k - s_j^j)$ so $\pi(s_k^k - s_j^j) = 0$.

11. $s = \pi(s_j^j)$ is a globally defined non-zero holomorphic section to $\text{coker}\{F\}$.

12. $\mathrm{coker}\{F\}$ is holomorphically trivial.

This establishes Assertion 1; Assertion 2 follows by duality. We now establish Assertion 3. We omit dz_i^i to define local sections to $\Lambda^{m,0}\mathbb{CP}^m$ over $\mathcal{O}_i$ by setting:

$$\omega_i := (-1)^i dz_i^1 \wedge \cdots \wedge dz_i^{i-1} \wedge dz_i^{i+1} \wedge \cdots \wedge dz_i^{m+1} .$$

We compute the transition functions of the canonical line bundle $\Lambda^{m,0}\mathbb{CP}^m$ on the intersection $\mathcal{O}_i \cap \mathcal{O}_j$; we take $i = 1$ and $j = 2$ for ease of computation. We have coordinates

$$\left(u^1 := \frac{z^2}{z^1}, u^2 := \frac{z^3}{z^1}, \ldots, u^m := \frac{z^{m+1}}{z^1} \right) \quad \text{on} \quad \mathcal{O}_1,$$

$$\left(v^1 := \frac{z^1}{z^2} = (u^1)^{-1}, v^2 := \frac{z^3}{z^2} = (u^1)^{-1}u^2, \ldots, v^m := \frac{z^m}{z^2} = (u^1)^{-1}u^m \right) \quad \text{on} \quad \mathcal{O}_2 .$$

Consequently,

$$\omega_1 = (-1)^1 du^1 \wedge du^2 \wedge \cdots \wedge du^m \quad \text{on} \quad \mathcal{O}_1,$$
$$\omega_2 = (-1)^2 dv^1 \wedge dv^2 \wedge \cdots \wedge dv^m \quad \text{on} \quad \mathcal{O}_2,$$
$$dv^1 = -(u^1)^{-2}du^1,$$
$$dv^2 = -(u^1)^{-2}u^2 du^1 + (u^1)^{-1}du^2, \ldots,$$
$$dv^m = -(u^1)^{-2}u^m + (u^1)^{-1}du^m,$$
$$\omega_2 = (-1)^1(u^1)^{-m-1}du^1 \wedge du^2 \wedge \cdots \wedge du^m = (u^1)^{-m-1}\omega_1.$$

By (4) above, $s_j = z_j^i s_i$ on $\mathcal{O}_i \cap \mathcal{O}_j$. Thus, in particular, $s_2 = (u^1)^{-1}s_1$ so the transition functions of $\Lambda^{m,0}\mathbb{CP}^m$ are $(u^1)^{-m-1}$. The third assertion now follows since the two line bundles have the same transition functions.

We now turn to the proof of the fourth assertion. We cover $\mathbb{L}$ by charts of the form $\mathcal{U}_i := \pi^{-1}\mathcal{O}_i$ where we take local coordinates $(\xi^i, \frac{z^1}{z^i}, \ldots, \frac{z^m}{z^i})$ where ξ^i is the fiber coordinate and where we omit $\frac{z^i}{z^i}$ as always. Do not sum over i and define

$$\Psi_i \left(\xi^i, \frac{z^1}{z^i}, \ldots, \frac{z^{m+1}}{z^i} \right) := \xi^i s_i \left(\frac{z^1}{z^i}, \ldots, \frac{z^{m+1}}{z^i} \right) .$$

Since $s_j = z_j^i s_i$, we have $\xi_j = (z_j^i)^{-1}\xi_i$. The remainder of the computation is the same as that which was performed to prove the third assertion where we must replace $(u^1)^{-m-1}$ by $(u^1)(u^1)^{-m-1} = (u^1)^{-m}$. □

By choosing a non-holomorphic splitting, we have the following vector bundle isomorphisms:

$$\mathbb{1} \oplus \Lambda^{1,0}(\mathbb{CP}^m) \approx \mathbb{L} \otimes \mathbb{1}^{m+1} \quad \text{and} \quad \mathbb{1} \oplus T_c(\mathbb{CP}^m) \approx \mathbb{L}^* \otimes \mathbb{1}^{m+1} . \tag{19.2.c}$$

These are not holomorphic isomorphisms. The isomorphism $\Lambda^{m,0}\mathbb{C}\mathbb{P}^m = \mathbb{L}^{m+1}$ is holomorphic. If M is a holomorphic manifold, then $\Lambda^{m,0}M$ is called the *canonical line bundle* of M.

In principal, one has to check the signs carefully in the proof of Assertion 3; we have not bothered because the ± 1 signs would define a flat real line bundle over $\mathbb{C}\mathbb{P}^m$; such a real line bundle would define a representation of the fundamental group of $\mathbb{C}\mathbb{P}^m$ into $\mathbb{Z}_2$. Since $\mathbb{C}\mathbb{P}^m$ is simply connected, such a representation is trivial and hence such a line bundle is in fact the flat trivial real line bundle. In principal, one could derive the third assertion from the first; we have given a separate argument since it is necessary to derive the fourth assertion. We have worked over $\mathbb{C}$. Let $\mathbb{L}_{\mathbb{R}}$ be the tautological or classifying real line bundle over $\mathbb{R}\mathbb{P}^m$. Exactly the same argument given to prove Lemma 19.9 yields that $T\mathbb{R}\mathbb{P}^m \oplus 1$ is isomorphic to $\mathbb{L}_{\mathbb{R}} \otimes 1^{m+1}$. We refer to the discussion in Example 18.20 for another more geometrical proof of this fact.

Theorem 19.10   We have $\int_{\mathbb{C}\mathbb{P}^1} c_1(\mathbb{L}) = -1$ and $\int_{\mathbb{C}\mathbb{P}^1} c_1(T_c\mathbb{C}\mathbb{P}^1) = 2$.

Proof. The Riemann sphere $\mathbb{C}\mathbb{P}^1 = S^2$ can be covered by two coordinate charts $(z, \mathbb{C})$ and $(w, \mathbb{C})$ where $w = \frac{1}{z}$. Let $s_1(z) := (1, z)$ and $s_2(w) := (w, 1)$ provide sections to $\mathbb{L}$ over $\mathcal{O}_1$ and $\mathcal{O}_2$, respectively. We give $\mathbb{C}^2$ the usual Euclidean inner product to define a Hermitian inner product $\mathbb{L}$. Since

$$h_1(z) := h(s_1(z), s_1(z)) = 1 + z\bar{z},$$

the connection one-form of ∇_h on $\mathcal{O}_1$ is given by $\omega_1 = \partial \log h_1 = (1 + z\bar{z})^{-1}\bar{z}\partial z$. We compute:

$$\begin{aligned}
c_1(\mathbb{L}) &= \tfrac{\sqrt{-1}}{2\pi}d\left\{(1 + z\bar{z})^{-1}\bar{z}\partial z\right\} = \tfrac{\sqrt{-1}}{2\pi}\left\{(1 + z\bar{z})^{-1} - (1 + z\bar{z})^{-2}z\bar{z}\right\}\bar{\partial}\bar{z} \wedge \partial z \\
&= \tfrac{\sqrt{-1}}{2\pi}(1 + x^2 + y^2)^{-2}(dx - \sqrt{-1}dy) \wedge (dx + \sqrt{-1}dy) \qquad\qquad (19.2.\text{d})\\
&= -\tfrac{1}{\pi}(1 + x^2 + y^2)^{-2}dx \wedge dy .
\end{aligned}$$

Since $\mathbb{C}\mathbb{P}^1 - \mathcal{O}_i$ consists of a single point, $\mathcal{O}_i$ is full measure in $\mathbb{C}\mathbb{P}^1$ and we have

$$\begin{aligned}
\int_{\mathbb{C}\mathbb{P}^1} c_1(\mathbb{L}) &= -\frac{1}{\pi}\int_{\mathbb{C}}(1 + x^2 + y^2)^{-2}dxdy = -\frac{1}{\pi}\int_0^\infty\int_0^{2\pi}(1 + r^2)^{-2}rd\theta dr \\
&= -2\int_0^\infty (1 + r^2)^{-2}rdr = (1 + r^2)^{-1}\big|_0^\infty = -1 .
\end{aligned}$$

We complete the proof by using Equation (19.2.c) to compute:

$$c_1(T_c\mathbb{C}\mathbb{P}^1) = -c_1(\Lambda^{1,0}\mathbb{C}\mathbb{P}^1) = -c_1(\Lambda^{1,0}\mathbb{C}\mathbb{P}^1 \oplus 1) = -c_1(\mathbb{L} \oplus \mathbb{L}) = -2c_1(\mathbb{L}) . \qquad \square$$

19.2.3 THE FUBINI–STUDY METRIC.

As noted previously in Section 19.2.1, there is a unique (up to scaling) $U(m + 1)$-invariant Riemannian metric on $\mathbb{C}\mathbb{P}^m$ which is called the *Fubini–Study metric*. We will now construct this metric quite concretely. Let $\mathbb{L}$ be the tautological line bundle over $\mathbb{C}\mathbb{P}^m$ discussed in Section 19.2.2. The canonical projection $\pi_2 : \mathbb{L} \to \mathbb{C}^{m+1}$ defines a canonical metric h on $\mathbb{L}$. We use the associated Chern form $-\pi c_1(\mathbb{L})$ to define a symmetric J-invariant bilinear form $g_{FS}(X, Y) = -\pi c_1(\mathbb{L})(X, JY)$.

Lemma 19.11 $U(m + 1)$ acts isometrically and transitively on $(\mathbb{CP}^m, g_{FS})$,

1. g_{FS} is given on the coordinate chart $\mathcal{O}_1$ by:

    (a) $g_{FS}(\partial_{x^i}, \partial_{x^i}) = g_{FS}(\partial_{y^i}, \partial_{y^i}) = (1 + \|\vec{z}\|^2 - z^i \bar{z}^i)(1 + \|\vec{z}\|^2)^{-2}$.

    (b) $g_{FS}(\partial_{x^i}, \partial_{y^i}) = 0$.

    (c) If $i \neq j$, then $g_{FS}(\partial_{x^i}, \partial_{x^j}) = g(\partial_{y^i}, \partial_{y^j}) = (1 + \|\vec{z}\|^2)^{-2}(-x^i x^j - y^i y^j)$.

    (d) If $i \neq j$, then $g_{FS}(\partial_{x^i}, \partial_{y^j}) = -g(\partial_{y^i}, \partial_{x^j}) = (1 + \|\vec{z}\|^2)^{-2}(y^i x^j - x^i y^j)$.

2. $(\mathbb{CP}^m, g_{FS})$ is a Kähler manifold.

3. Let $\Psi(r, \theta) := \tan(r)\theta$ for $r \in [0, \frac{\pi}{2})$ and $\theta \in S^{2m-1} \subset \mathbb{C}^m = \mathcal{O}_1$.

    (a) The curves $r \to \Psi(r, \theta)$ are unit speed geodesics in $\mathbb{CP}^m$ for any θ.

    (b) Ψ defines geodesic polar coordinates on $\mathbb{CP}^m$.

    (c) $\operatorname{diam}(\mathbb{CP}^m) = \frac{\pi}{2}$ and $\operatorname{vol}(\mathbb{CP}^m) = \frac{\pi^m}{m!}$.

Proof. Since J and $c_1(\mathbb{L})$ are invariant under the action of $U(m + 1)$, g_{FS} is invariant; since $U(m + 1)$ acts transitively on S^{2m+1}, $U(m + 1)$ acts transitively on $\mathbb{CP}^m$. Let s be the canonical section to $\mathbb{L}$ over $\mathcal{O}_1$ given in Equation (19.2.a); $h(s, s) = 1 + \|\vec{z}\|^2$. Consequently, Equation (18.5.d) yields

$$
\begin{aligned}
-\pi c_1(L, h) &= -\frac{\sqrt{-1}}{2} \bar{\partial} \left\{ (1 + \|\vec{z}\|^2)^{-1} \bar{z}^i dz^i \right\} \\
&= \frac{\sqrt{-1}}{2} \left\{ -(1 + \|\vec{z}\|^2)^{-2} z^i \bar{z}^j dz^j \wedge d\bar{z}^i + (1 + \|\vec{z}\|^2)^{-1} dz^i \wedge d\bar{z}^i \right\} .
\end{aligned} \tag{19.2.e}
$$

Let $z^i = x^i + \sqrt{-1} y^i$. We rewrite Equation (19.2.e) in the form

$$
\begin{aligned}
-\pi c_1(L, h) \quad = \quad & (1 + \|\vec{z}\|^2)^{-2} \sum_i (1 + \|\vec{z}\|^2 - |z^i|^2) dx^i \wedge dy^i \\
& + (1 + \|\vec{z}\|^2)^{-2} \sum_{i<j} \{ (x^i x^j + y^i y^j)(dy^i \wedge dx^j - dx^i \wedge dy^j) \} \\
& + (1 + \|\vec{z}\|^2)^{-2} \sum_{i<j} \{ (x^i y^j - x^j y^i)(dx^i \wedge dx^j + dy^i \wedge dy^j) \} .
\end{aligned}
$$

Assertion 1 now follows. We set $\vec{z} = 0$ to see that g_{FS} is positive definite at the origin. Since $U(m + 1)$ acts transitively on $\mathbb{CP}^m$ and g_{FS} is invariant, g_{FS} is positive definite on all of $\mathbb{CP}^m$. Consequently, g_{FS} defines a Hermitian metric on $\mathbb{CP}^m$. The associated Kähler form is $-\pi c_1(\mathbb{L}, h)$. As $dc_1(\mathbb{L}, h) = 0$, g_{FS} is Kähler; Assertion 2 follows. Let P be a point of $\mathbb{CP}^m$. Since $\mathbb{CP}^m$ is a homogeneous space, we may assume without loss of generality that $P = \langle 1, 0, \dots, 0 \rangle$ in studying the geometry about P. Let $\sigma(t) = (t, 0, \dots, 0)$ so $\dot{\sigma} = \partial_{x^1}$. By Lemma 16.2,

$$
\begin{aligned}
g_{FS}(\nabla_{\dot{\sigma}} \dot{\sigma}, \partial_{x^j}) &= \tfrac{1}{2} \{ 2\partial_{x^1} g_{FS}(\partial_{x^1}, \partial_{x^j}) - \partial_{x^j} g_{FS}(\partial_{x^1}, \partial_{x^1}) \}(\sigma), \\
g_{FS}(\nabla_{\dot{\sigma}} \dot{\sigma}, \partial_{y^j}) &= \tfrac{1}{2} \{ 2\partial_{x^1} g_{FS}(\partial_{x^1}, \partial_{y^j}) - \partial_{y^j} g_{FS}(\partial_{x^1}, \partial_{x^1}) \}(\sigma) .
\end{aligned} \tag{19.2.f}
$$

By Assertion 1, $g_{FS}(\partial_{x^1}, \partial_{x^1}) = (1 + \|\vec{z}\|^2 - \|z^1\|^2)(1 + \|\vec{z}\|^2)^{-2}$. Since $x^j(\sigma) = 0$ for $j > 1$ and $y^j(\sigma) = 0$ for any j, we have

$$\partial_{x^j} g_{FS}(\partial_{x^1}, \partial_{x^1})(\sigma) = 0 \text{ for } j > 1, \quad \text{and} \quad \partial_{y^j} g_{FS}(\partial_{x^1}, \partial_{x^1})(\sigma) = 0 \text{ for any } j. \qquad (19.2.g)$$

We may also use Assertion 1 to compute similarly that

$$\partial_{x^1} g_{FS}(\partial_{x^1}, \partial_{x^j})\}(\sigma) = 0 \text{ for } j > 1, \quad \text{and} \quad \partial_{x^1} g_{FS}(\partial_{x^1}, \partial_{y^j})\}(\sigma) = 0 \text{ for any } j. \qquad (19.2.h)$$

We use Equations (19.2.f), (19.2.g), and (19.2.h) to conclude

$$g_{FS}(\nabla_{\dot\sigma}\dot\sigma, \partial_{x^j})(\sigma) = 0 \text{ for } j > 1, \quad \text{and} \quad g_{FS}(\nabla_{\dot\sigma}\dot\sigma, \partial_{y^j})(\sigma) = 0 \text{ for any } j . \qquad (19.2.i)$$

By Assertion 1, $\{\partial_{x^i}, \partial_{y^j}\}$ is an orthogonal frame along σ. Consequently, Equation (19.2.i) shows that $\nabla_{\dot\sigma}\dot\sigma$ is a multiple of $\dot\sigma$ so σ is an unparameterized geodesic. We have

$$g_{FS}(\dot\sigma, \dot\sigma) = (1 + t^2)^{-2} .$$

Let $\tau(r) = (\tan(r), 0, \ldots, 0)$. Since $\tan' = 1 + \tan^2$, $\|\dot\tau\|^2 = 1$ so τ is a unit speed geodesic. If $\theta \in S^{2m-1}$, let $\tau_\theta(r) = \tan(r)\theta$. Choose $\Phi \in U(m)$ so $\Phi(1, 0, \ldots, 0) = \theta$. Since $\tau_\theta = \Phi\tau$, τ_θ is a unit speed geodesic. This shows Ψ defines geodesic polar coordinates on $\mathbb{CP}^m$. The map $\Psi_1 : \vec{z} \to \langle 1, \vec{z}\rangle$ embeds $\mathcal{O}_1$ in $\mathbb{CP}^m$. For $\lambda \in S^1$, let $\sigma_\lambda(r) := (\lambda \tan(r), 0, \ldots, 0)$ be a unit speed geodesic in $\mathbb{CP}^m$.

$$\Psi_1 \circ \sigma_\lambda(r) = \langle 1, \tan(r), 0, \ldots, 0\rangle = \langle \lambda^{-1} \cot(r), 1, 0, \ldots, 0\rangle \quad \text{if} \quad 0 < |r| < \tfrac{\pi}{2},$$

$$\lim_{r \to \pm\frac{\pi}{2}} \Psi_1 \circ \sigma_\lambda(r) = \langle 0, 1, 0, \ldots, 0\rangle \in \mathbb{CP}^m - \mathcal{O}_1 = \mathbb{CP}^{m-1} .$$

Thus, there there is a family of distinct distance minimizing geodesics parametrized by $\lambda \in S^1$ between $\langle 1, 0, 0 \ldots, 0\rangle$ and $\langle 0, 1, 0, \ldots, 0\rangle$ all of which have length $\frac{\pi}{2}$. This shows the cut locus of $\langle 1, 0, \ldots, 0\rangle$ is the hypersurface $\mathbb{CP}^m - \mathcal{O}_1 = \mathbb{CP}^{m-1}$ and the diameter of $\mathbb{CP}^m$ is $\frac{\pi}{2}$.

The non-zero components of the Fubini–Study metric on $\sigma(t)$ are

$$g_{FS}(\partial_{x^1}, \partial_{x^1})(\sigma(t)) = g_{FS}(\partial_{y^i}, \partial_{y^i})(\sigma(t)) = (1 + t^2)^{-2},$$

$$g_{FS}(\partial_{x^j}, \partial_{x^j})(\sigma(t)) = g_{FS}(\partial_{y^j}, \partial_{y^j})(\sigma(t)) = (1 + t^2)^{-1} .$$

This shows that $\mathrm{dvol}_{g_{FS}}(r, 0, \ldots, 0) = (1 + t^2)^{-m-1} \mathrm{dvol}_{\mathbb{C}^m}$. Using invariance under the unitary group we see $\mathrm{dvol}_{g_{FS}} = (1 + \|\vec{z}\|^2)^{-m-1} \mathrm{dvol}_{\mathbb{C}^m}$. Changing to rectangular polar coordinates (t, θ) yields $\mathrm{dvol}_{g_{FS}} = t^{2m-1}(1 + t^2)^{-m-1} dt \, \mathrm{dvol}_{S^{2m-1}}$. Set $r = \tan(t)$ to obtain

$$\mathrm{dvol}_{g_{FS}} = \sin(r)^{2m-1} \cos(r) dr \, \mathrm{dvol}_{S^{2m-1}}$$

in geodesic polar coordinates. Consequently, the *volume density function* of $\mathbb{CP}^m$ is given by $\sin(r)^{2m-1} \cos(r)$; this was first established by Watanabe [74]. Since

$$\mathrm{vol}(S^{2m-1}) = 2\pi^m/(m-1)!,$$

we obtain

$$\text{vol}(\mathbb{CP}^m) = 2\pi^m/(m-1)! \cdot \int_{r=0}^{\frac{\pi}{2}} \sin(r)^{2m-1} \cos(r) dr = \pi^m/m! \,. \qquad \square$$

We introduce the following tensors to discuss the curvature of g_{FS}:

$$R_0(\xi_1, \xi_2, \xi_3, \xi_4) := g(\xi_1, \xi_4) g(\xi_2, \xi_3) - g(\xi_1, \xi_3) g(\xi_2, \xi_4),$$
$$R_J(\xi_1, \xi_2, \xi_3, \xi_4) := g(\xi_1, J\xi_4) g(\xi_2, J\xi_3) - g(\xi_1, J\xi_3) g(\xi_2, J\xi_4)$$
$$- 2g(\xi_1, J\xi_2) g(\xi_3, J\xi_4).$$

These are *algebraic curvature tensors*; both R_0 and R_J satisfy the first Bianchi identity and the usual $\mathbb{Z}_2$ symmetries of the curvature tensor. These two tensors play an important role in the study of Osserman geometry; the tensor R_0 is the algebraic curvature tensor of constant sectional curvature $+1$; we refer to [27, 53–56] for further details.

Lemma 19.12 Let $R(X, Y, Z, W)$ be the curvature tensor of $(\mathbb{CP}^m, g_{FS})$.

1. Up to the usual $\mathbb{Z}_2$ symmetries, the non-zero components of R are

   (a) $R(\partial_{x^i}, \partial_{y^i}, \partial_{y^i}, \partial_{x^i})(0) = 4$.

   (b) $R(\partial_{x^i}, \partial_{x^j}, \partial_{x^j}, \partial_{x^i})(0) = R(\partial_{x^i}, \partial_{y^j}, \partial_{y^j}, \partial_{x^i})(0) = R(\partial_{y^i}, \partial_{y^j}, \partial_{y^j}, \partial_{y^i})(0) = 1$.

   (c) $R(\partial_{x^i}, \partial_{x^j}, \partial_{y^j}, \partial_{y^i})(0) = -R(\partial_{x^i}, \partial_{y^j}, \partial_{x^j}, \partial_{y^i})(0) = 1$

   (d) $R(\partial_{x^i}, \partial_{y^i}, \partial_{x^j}, \partial_{y^j})(0) = -2$.

2. We have $R = R_0 + R_J$ and $\nabla R = 0$.

Proof. We use Lemma 19.11. Since $U(m + 1)$ acts transitively on $\mathbb{CP}^m$, it suffices to compute at the origin of the coordinate system $\mathcal{O}_1$, i.e., we may set $\vec{z} = 0$. By Lemma 19.11,

$$g_{ij} = \delta_{ij} + O(\|\vec{z}\|^2) \,.$$

We may therefore apply Lemma 16.2 to see:

$$R_{ijk\ell}(0) = \tfrac{1}{2}\{g_{j\ell/ik} + g_{jk/i\ell} - g_{i\ell/jk} - g_{jk/i\ell}\} \,. \qquad (19.2.\text{j})$$

Let ξ_i be coordinate vector fields. We apply Equation (19.2.j) and Lemma 19.11 to see that the only possibly non-zero curvatures $R(\xi_1, \xi_2, \xi_3, \xi_4)(0)$ occur when the set $S := \{\xi_1, \xi_2, \xi_3, \xi_4\}$ consists of just 2 distinct coordinate vector fields or when the ξ_i are a permutation of the coordinate vector fields $\{\partial_{x^i}, \partial_{y^i}, \partial_{x^j}, \partial_{y^j}\}$ for suitably chosen $i \neq j$.

We expand $(1 + \|\vec{z}\|^2)^{-2} = 1 - 2\|\vec{z}\|^2 + O(\|\vec{z}\|^4)$ to see

$$\begin{aligned}
g(\partial_{x^i}, \partial_{x^i}) &= (1 - 2\|\vec{z}\|)(1 + \|\vec{z}\|^2 - x^i x^i - y^i y^i) + O(\|\vec{z}\|^4) \\
&= 1 - \sum_{j \neq i}(x^j x^j + y^j y^j) - 2(x^i x^i + y^i y^i) + O(\|\vec{z}\|^4),
\end{aligned}$$

$$R(\partial_{x^i}, \partial_{y^i}, \partial_{y^i}, \partial_{x^i})(0) = -\tfrac{1}{2}\{\partial_{y^i}\partial_{y^i} g(\partial_{x^i}, \partial_{x^i}) + \partial_{x^i}\partial_{x^i} g(\partial_{y^i}, \partial_{y^i})\} = 4\,;$$

Assertion 1a follows. We compute:

$$\begin{aligned}
R(\partial_{x^i}, \partial_{x^j}, \partial_{x^j}, \partial_{x^i}) &= \tfrac{1}{2}\{2\partial_{x^i}\partial_{x^j}g(\partial_{x^i}, \partial_{x^j}) - \partial_{x^i}\partial_{x^i}g(\partial_{x^j}, \partial_{x^j}) - \partial_{x^j}\partial_{x^j}g(\partial_{x^i}, \partial_{x^i})\} \\
&= \tfrac{1}{2}\{-2 + 2 + 2\} = 1.
\end{aligned}$$

We can use the action of $U(m+1)$ to replace ∂_{x^i} by ∂_{y^i} or to replace ∂_{x^j} by ∂_{y^j} and complete the proof of Assertion 1b. Since the Fubini–Study metric is Kähler, Theorem 18.37 yields the identity $R(x, y, z, w) = R(Jx, Jy, z, w) = R(x, y, Jz, Jw)$. It now follows $R(\partial_{x^i}, \partial_{x^j}, \partial_{y^j}, \partial_{y^i})(0) = 1$. We can find $\Psi \in U(m+1)$ so

$$\Psi_*\partial_{x^i} = \partial_{x^i}, \quad \Psi_*\partial_{y^i} = \partial_{y^i}, \quad \Psi_*\partial_{x^j} = \partial_{y^j}, \quad \Psi_*\partial_{y^j} = -\partial_{x^j}.$$

Since $\Psi^* R = R$, we have $R(\partial_{x^i}, \partial_{y^j}, \partial_{x^j}, \partial_{y^i}) = R(\partial_{x^i}, -\partial_{x^j}, \partial_{y^j}, \partial_{y^i}) = -1$. This verifies Assertion 1c. We use the first Bianchi identity to see

$$\begin{aligned}
0 &= R(\partial_{x^i}, \partial_{y^i}, \partial_{x^j}, \partial_{y^j}) + R(\partial_{x^i}, \partial_{x^j}, \partial_{y^j}, \partial_{y^i}) + R(\partial_{x^i}, \partial_{y^j}, \partial_{y^i}, \partial_{x^j}) \\
&= R(\partial_{x^i}, \partial_{y^i}, \partial_{x^j}, \partial_{y^j}) + 1 + 1.
\end{aligned}$$

This verifies Assertion 1d and completes the proof of Assertion 1. We perform a direct computation to see $R = R_0 + R_J$. Since the linear and cubic terms in the expansion of g around the origin vanish, it follows that $\nabla R = 0$. □

We use *stereographic projection* (see Section 3.6.5 of Book I) to parametrize the punctured sphere $S^2 - (0, 0, 1)$ by setting

$$T(x, y) := (1 + x^2 + y^2)^{-1}(2x, 2y, x^2 + y^2 - 1).$$

We then have $ds_{S^2}^2 = 4(1 + x^2 + y^2)^{-2}(dx^2 + dy^2)$. By Equation (19.2.d),

$$ds_{FS}^2 = (1 + x^2 + y^2)^{-2}(dx^2 + dy^2) \quad \text{on} \quad \mathcal{O}_1 \subset \mathbb{CP}^1.$$

Thus, $ds_{FS}^2 = \tfrac{1}{4}ds_{S^2}^2$; $(\mathbb{CP}^1, g_{FS})$ is the sphere of radius $\tfrac{1}{2}$ and, therefore, has sectional curvature 4 rather than 1. This gives another proof that $\mathrm{diam}(\mathbb{CP}^1, g_{FS}) = \tfrac{\pi}{2}$ and $\mathrm{vol}(\mathbb{CP}^1, g_{FS}) = \pi$. Let S^1 and S^{2m+1} be the circle and sphere of radius 1. The *Hopf fibration* $S^1 \to S^{2m+1} \to \mathbb{CP}^m$ is a Riemannian submersion and we have a second verification that

$$\mathrm{vol}(\mathbb{CP}^m) = \frac{\mathrm{vol}(S^{2m+1})}{\mathrm{vol}(S^1)} = \frac{2\pi^{m+1}/m!}{2\pi} = \frac{\pi^m}{m!}.$$

19.3 HODGE MANIFOLDS

Hodge manifolds are defined in Section 19.3.1 and integrality conditions are discussed in Section 19.3.2. The Riemann period relations are used in Section 19.3.3 to provide examples of flat tori which are not Hodge manifolds. In Section 19.3.4, we discuss the Kähler potential.

19.3.1 THE BASIC PROPERTIES OF HODGE MANIFOLDS. Let h be a Hermitian inner product on a holomorphic line bundle L over a complex manifold M. Let

$$c_1(L, h) = \frac{\sqrt{-1}}{2\pi} \bar{\partial}\partial \log h$$

be the first Chern form and let $g_h(X, Y) := c_1(L, h)(X, JY)$ be the associated symmetric bilinear form. We say that (L, h) is a *positive line bundle* if g_h is positive definite; the associated Kähler form is then given by $c_1(L, h)$. By an abuse of notation, we will say that L is positive or that L is a positive line bundle if there exists a fiber metric so that (L, h) is positive. In this setting, we say that g_h is Hodge.

Lemma 19.13   Let (L, h) be a positive line bundle and let (L_1, h_1) be an arbitrary Hermitian holomorphic line bundle over a compact holomorphic manifold (M, J).

1. If $L \otimes (\Lambda^{m,0} M)^*$ is a positive line bundle, then $H^{0,q}(M, L) = 0$ for $q > 0$.

2. There exists $k_0 = k_0(L, L_1, h, h_1)$ so if $k \geq k_0$, then $L^k \otimes L_1$ is a positive line bundle.

Proof. By Theorem 18.41, if $L \otimes (\Lambda^{m,0} M)^*$ is a positive line bundle and if $p + q > \mathfrak{m}$, then $H^{p,q}(M, L \otimes (\Lambda^{m,0} M)^*) = 0$. We set $p = \mathfrak{m}$. If $q > 0$, Assertion 1 follows because

$$H^{\mathfrak{m},q}(M, L \otimes (\Lambda^{m,0} M)^*) = H^{0,q}(M, L \otimes (\Lambda^{m,0} M)^* \otimes \Lambda^{m,0} M) = H^{0,q}(M, L).$$

If (L, h) is a positive line bundle and if (L_1, h_1) is arbitrary, then Equation (18.5.e) yields

$$c_1(L^k \otimes L_1, h^k \otimes h_1) = k \cdot c_1(L, h) + c_1(L_1, h_1),$$
$$g_{(L^k \otimes L_1, h^k \otimes h_1)} = k g_{(L,h)} + g_{(L_1, h_1)}.$$

Consequently, $\lim_{k \to \infty} \frac{1}{k} g_{(L^k \otimes L_1, h^k \otimes h_1)} = g_{(L,h)}$. The set of positive definite bilinear Hermitian forms is an open subset of the set of all bilinear Hermitian forms on the tangent bundle. Since M is compact, there exists k so $\frac{1}{k} g_{(L^k \otimes L_1, h^k \otimes h_1)}$ is positive definite and hence $g_{(L^k \otimes L_1, h^k \otimes h_1)}$ is positive definite. This proves Assertion 2.   □

If (M, J) admits a positive line bundle, we say that (M, J) is a *Hodge manifold*.

Lemma 19.14   Let (M, g, J) be a Hermitian manifold.

1. If (M, J) is a Hodge manifold, then (M, J) admits a Kähler metric.

2. Let $(N, g|_N, J|_N)$ be a Hermitian submanifold of (M, g, J).

   (a) If (M, g, J) is Kähler, then $(N, g|_N, J|_N)$ is Kähler.

   (b) If (M, J) is a Hodge manifold, then $(N, J|_N)$ is a Hodge manifold.

3. $(\mathbb{CP}^m, J)$ is a Hodge manifold.

4. The Cartesian product of Hodge manifolds is again a Hodge manifold.

5. If $\mathfrak{m} = 1$, then (M, g, J) is a Hodge manifold.

Proof. If (M, J) is a Hodge manifold, then $d\Omega_{g_h} = dc_1(L, h) = 0$ and (M, J) admits a Kähler metric. Let i be the inclusion of N in M. If (M, J, g) is Kähler, then we have that $d\Omega_N = di^*\Omega_M = i^*d\Omega_M = 0$ and $\mathcal{N}$ is Kähler. If (L, h) is a positive line bundle over (M, J), then $c_1(i^*L, i^*h) = i^*c_1(L, h)$ so (i^*L, i^*h) is a positive line bundle over N and (N, J_N) is a Hodge manifold. Because $\frac{1}{\pi}g_{FS} = c_1(\mathbb{L}^*)$, $\mathbb{L}^*$ is a positive line bundle over $\mathbb{CP}^{\mathfrak{m}}$ and $\mathbb{CP}^{\mathfrak{m}}$ is a Hodge manifold. Since $c_1(L_1 \otimes L_2, h_1 \otimes h_2) = c_1(L_1, h_1)c_1(L_2, h_2)$, if (L_i, h_i) are positive line bundles over holomorphic manifolds M_i, then we have that $(L_1 \otimes L_2, h_1 \otimes h_2)$ is a positive line bundle over $M_1 \times M_2$. $\square$

19.3.2 INTEGRALITY CONDITIONS.

The condition that (M, J) is a Hodge manifold is an algebraic one. Thus, the question of finding a positive line bundle is related to finding an element of $H^2(M; \mathbb{Z}) \cap H^{1,1}(M; \mathbb{C})$ which is positive. This is not always possible. Let $f(z^1, \ldots, z^{\mathfrak{m}+1})$ be a non-trivial homogeneous polynomial of degree k in $\mathfrak{m} + 1$ complex variables. Since $f(\lambda z^1, \ldots, \lambda z^{\mathfrak{m}+1}) = \lambda^k f(z^1, \ldots, z^{\mathfrak{m}+1})$, the zero set

$$Z_f := \{\langle \vec{z} \rangle \in \mathbb{CP}^{\mathfrak{m}} : f(\vec{z}) = 0\} \subset \mathbb{CP}^{\mathfrak{m}}$$

is well defined. Let $f_i(\vec{w}) = f(w_1, \ldots, w_{i-1}, 1, w_{i+1}, \ldots, w_{\mathfrak{m}})$ on $\mathcal{O}_i$. If we assume that the complex gradient $\mathrm{grad}^c(f_i) \neq 0$ on $Z_f \cap \mathcal{O}_i$, then Z_f is a holomorphic submanifold of $\mathbb{CP}^{\mathfrak{m}}$.

Remark 19.15 The zero set of the function $f(z) = w_1^4 + w_2^4 + w_3^4 + w_4^4 = 0$ defines a *Fermat surface* $\mathcal{S} \subset \mathbb{CP}^3$. This surface is a spin manifold which does not admit a metric of positive scalar curvature (see Remark 18.24). The surface $\mathcal{S}$ is a K3 *surface* (named by Weil [76] after Kummer, Kähler and Kodaira: "Dans la seconde partie de mon rapport, il ságit des variétés kählériennes dites K3, ainsi nommées en l'honneur de Kummer, Kähler, Kodaira et de la belle montagne K2 au Cachemire". The *canonical line bundle* $\Lambda^{2,0}(\mathcal{S})$ is trivial. Let $\beta_{p,q} := \dim(H^{p,q})$. Then

$$\beta_{0,0} = \beta_{2,2} = \beta_{2,0} = \beta_{0,2} = 1 \quad \text{and} \quad \beta_{1,1} = 20.$$

The remaining Betti numbers vanish. *Kummer surfaces* [45] are K3 surfaces which are defined projectively by the equation

$$
\begin{aligned}
0 = {} & (x^2 + y^2 + z^2 - \mu^2 w^2)^2 \\
& -\lambda(w - z - \sqrt{2}x)(w - z + \sqrt{2}x)(w + z + \sqrt{2}y)(w + z - \sqrt{2}y) = 0
\end{aligned}
$$

(see [48]). Any K3 surface admits a Kähler metric but not all K3 surfaces are Hodge manifolds. The moduli space of K3 surfaces is 20-dimensional; the moduli space of K3 surfaces which are Hodge is 19-dimensional (see [78]).

19.3.3 THE RIEMANN PERIOD RELATIONS. Let $\mathbb{T}^4 := S^1 \times S^1 \times S^1 \times S^1$ with the usual periodic parameters x^1, x^2, x^3, and x^4. We may identify $\mathbb{T}^4$ with $\mathbb{R}^4/\mathbb{Z}^4$. Let ε be a real parameter. Define a holomorphic structure on $\mathbb{T}^4$ so that

$$z^1 = x^1 + \sqrt{-1}x^2 \quad \text{and} \quad z^2 = x^3 + \sqrt{-1}(x^4 + \varepsilon x^2)$$

are local holomorphic coordinates for $\mathbb{T}^4$. We have

$$dz^1 = dx^1 + \sqrt{-1}dx^2, \qquad\qquad dz^2 = dx^3 + \sqrt{-1}(dx^4 + \varepsilon dx^2),$$
$$\partial_{z^1} = \tfrac{1}{2}\{\partial_{x^1} - \sqrt{-1}(\partial_{x^2} - \varepsilon\partial_{x^4})\}, \quad \partial_{z^2} = \tfrac{1}{2}(\partial_{x^3} - \sqrt{-1}\partial_{x^4}).$$

The complex structure is given by

$$J_\varepsilon^* dx^1 = -dx^2, \qquad\qquad J_\varepsilon^* dx^2 = dx^1,$$
$$J_\varepsilon^* dx^3 = -dx^4 - \varepsilon dx^2, \quad J_\varepsilon^* dx^4 = dx^3 - \varepsilon dx^2,$$
$$J_\varepsilon \partial_{x^1} = \partial_{x^2}, \qquad\qquad J_\varepsilon \partial_{x^2} = -\partial_{x^1} - \varepsilon\partial_3 - \varepsilon\partial_4,$$
$$J_\varepsilon \partial_{x^3} = \partial_{x^4}, \qquad\qquad J_\varepsilon \partial_{x^4} = -\partial_{x^3}.$$

We define a Kähler form and associated Kähler metric by setting

$$\Omega = \tfrac{\sqrt{-1}}{2}dz^1 \wedge d\bar{z}^2 + dz^2 \wedge d\bar{z}^2 = dx^1 \wedge dx^2 + dx^3 \wedge (dx^4 + \varepsilon dx^2),$$
$$g = (dz^1 \circ d\bar{z}^1) + (dz^2 \circ d\bar{z}^2)$$
$$= dx^1 \circ dx^1 + dx^2 \circ dx^2 + dx^3 \circ dx^3 + (dx^4 + \varepsilon dx^2) \circ (dx^4 + \varepsilon dx^2).$$

Let $\mathcal{T}_\varepsilon := (\mathbb{T}^2, J_\varepsilon)$ be the resulting holomorphic manifold. We define a flat Kähler metric on $\mathcal{T}_\varepsilon$ by setting

$$\Omega_\varepsilon := dz^1 \wedge d\bar{z}^1 + dz^2 \wedge d\bar{z}^2 = -2\sqrt{-1}(dx^1 \wedge dx^2 + dx^3 \wedge dx^4 + \varepsilon dx^3 \wedge dx^2),$$
$$g_\varepsilon := dz^1 \circ d\bar{z}^1 + dz^2 \circ d\bar{z}^2 = (dx^1)^2 + (dx^2)^2 + (dx^3)^2 + (dx^4 + \varepsilon dx^2)^2.$$

Lemma 19.16

1. If $\varepsilon \in \mathbb{R} - \mathbb{Q}$, then $\mathcal{T}_\varepsilon$ is not a Hodge manifold.
2. If $\varepsilon = 0$, then $\mathcal{T}_\varepsilon$ is a Hodge manifold.

Proof. Let ε be irrational. Assume L is a holomorphic line bundle over $\mathcal{T}_\varepsilon$ with $c_1(L)$ positive. We argue for a contradiction. Since $c_1(L)$ is a real form of Type (1,1), there are smooth coefficients so

$$c_1(L) = \tfrac{\sqrt{-1}}{4\pi^2}\{a_{1\bar{1}}dz^1 \wedge d\bar{z}^1 + a_{1\bar{2}}dz^1 \wedge d\bar{z}^2 + a_{2\bar{1}}dz^2 \wedge d\bar{z}^1 + a_{2\bar{2}}dz^2 \wedge d\bar{z}^2\}.$$

The normalizing constant $4\pi^2$ is for later convenience. Since $c_1(L)$ is positive, $a_{2\bar{2}}$ is positive. We pass to de Rham cohomology;

$$H^2(\mathbb{T}^4;\mathbb{C}) = \text{span}\{[dz^1 \wedge dz^2], [d\bar{z}^1 \wedge d\bar{z}^2], [dz^1 \wedge d\bar{z}^1],$$

$$[dz^1 \wedge d\bar{z}^2], [dz^2 \wedge d\bar{z}^1], [dz^2 \wedge d\bar{z}^1]\}.$$

We may therefore express

$$[c_1(L)] = \tfrac{\sqrt{-1}}{4\pi^2}\{A_{1\bar{1}}[dz^1 \wedge d\bar{z}^1] + A_{1\bar{2}}[dz^1 \wedge d\bar{z}^2] + A_{2\bar{1}}[dz^2 \wedge d\bar{z}^1] + A_{2\bar{2}}[dz^2 \wedge d\bar{z}^2]\}$$

where the constants $A_{\alpha\bar{\beta}}$ are obtained by averaging the $a_{\alpha\bar{\beta}}$ over $\mathbb{T}^4$ to obtain the corresponding Fourier coefficients. In particular, $A_{2\bar{2}}$ is positive.

1. Let $\phi_1(u, v) = (0, u, v, 0)$. Then

$$dz^1 = dx^1 + \sqrt{-1}dx^2, \qquad\qquad dz^2 = dx^3 + \sqrt{-1}(dx^4 + \varepsilon dx^2),$$
$$\phi_1^*(dz^1) = \sqrt{-1}du, \qquad\qquad \phi_1^*(dz^2) = dv + \sqrt{-1}\varepsilon du,$$
$$\phi_1^*(dz^1 \wedge d\bar{z}^1) = 0, \qquad\qquad \phi_1^*(dz^1 \wedge d\bar{z}^2) = \sqrt{-1}du \wedge dv,$$
$$\phi_1^*(dz^2 \wedge d\bar{z}^1) = -\sqrt{-1}dv \wedge du, \qquad \phi_1^*(dz^2 \wedge d\bar{z}^2) = -2\varepsilon\sqrt{-1}dv \wedge du,$$
$$\int_{\mathbb{T}^2} \phi_1^* c_1(L) = -A_{1\bar{2}} - A_{2\bar{1}} - 2\varepsilon A_{2\bar{2}}.$$

2. Let $\phi_2(u, v) = (u, 0, 0, v)$. Then

$$dz^1 = dx^1 + \sqrt{-1}dx^2, \qquad\qquad dz^2 = dx^3 + \sqrt{-1}(dx^4 + \varepsilon dx^2),$$
$$\phi_2^*(dz^1) = du, \qquad\qquad\qquad \phi_2^*(dz^2) = \sqrt{-1}dv,$$
$$\phi_2^*(dz^1 \wedge d\bar{z}^1) = 0, \qquad\qquad \phi_2^*(dz^1 \wedge d\bar{z}^2) = -\sqrt{-1}du \wedge dv,$$
$$\phi_2^*(dz^2 \wedge d\bar{z}^1) = \sqrt{-1}dv \wedge du, \quad \phi_2^*(dz^2 \wedge d\bar{z}^2) = 0,$$
$$\int_{\mathbb{T}^2} \phi_2^* c_1(L) = -A_{1\bar{2}} - A_{2\bar{1}}.$$

3. Let $\phi_3(u, v) = (0, 0, u, v)$.

$$dz^1 = dx^1 + \sqrt{-1}dx^2, \quad dz^2 = dx^3 + \sqrt{-1}(dx^4 + \varepsilon dx^2),$$
$$\phi_3^*(dz^1) = 0, \qquad\qquad \phi_3^*(dz^2) = du + \sqrt{-1}dv,$$
$$\phi_3^*(dz^1 \wedge d\bar{z}^1) = 0, \qquad \phi_3^*(dz^1 \wedge d\bar{z}^2) = 0,$$
$$\phi_3^*(dz^2 \wedge d\bar{z}^1) = 0, \qquad \phi_3^*(dz^2 \wedge d\bar{z}^2) = -2\sqrt{-1}du \wedge dv,$$
$$\int_{\mathbb{T}^2} \phi_3^* c_1(L) = -2A_{2\bar{2}}.$$

By Lemma 16.11, $\int_{\mathbb{T}^2} \phi_i^* c_1(L) = \int_{\mathbb{T}^2} c_1(\phi_i^* L) \in \mathbb{Z}$. Consequently,

$$-A_{1\bar{2}} - A_{2\bar{1}} - 2\varepsilon A_{2\bar{2}} \in \mathbb{Z}, \quad -A_{1\bar{2}} - A_{2\bar{1}} \in \mathbb{Z}, \quad A_{2\bar{2}} \in \mathbb{Z}.$$

This implies that $-2\varepsilon A_{2\bar{2}} \in \mathbb{Z}$ and that $-2A_{2\bar{2}} \in \mathbb{Z}$. Since ε is irrational, we conclude that $A_{2\bar{2}} = 0$ which is false. On the other hand, $\mathbb{T}_0^4$ is the product torus $\mathbb{T}^2 \times \mathbb{T}^2$. Assertion 4 of Lemma 19.14 implies $\mathbb{T}^2$ is a Hodge manifold; Assertion 5 of Lemma 19.14 then implies the product $\mathbb{T}_0^4 = \mathbb{T}^2 \times \mathbb{T}^2$ is a Hodge manifold.   □

We have chosen to fix the lattice and vary the complex structure. Alternatively, of course, we could have regarded $\mathbb{T}_\varepsilon$ as the quotient of $\mathbb{C}^2$ by a lattice Γ_ε of rank 4 in $\mathbb{C}^2$. The resulting algebraic conditions that a general torus $\mathbb{C}^m/\Gamma$ is a Hodge manifold are then referred to as the *Riemann period relations*. We refer to the discussion in Birkenhake and Lange [9], Griffiths and Harris [31], Gunning [34], and Riemann [63] for further details.

19.3.4 THE KÄHLER POTENTIAL. We can use Lemma 19.7 (the Dolbeault Lemma) to show that every Kähler metric is a Hodge metric locally. The obstruction to a Hodge structure is a global one.

Lemma 19.17   Let $B_{\vec{\varepsilon}}(\vec{z}_0) \subset \mathbb{C}^m$ be a poly-disk.

1. If Ω is a real two-form of Type (1,1) on $B_{\vec{\varepsilon}}(\vec{z}_0)$ with $d\Omega = 0$, then there exists a smooth real-valued function ϕ on $B_{\vec{\varepsilon}}(\vec{z}_0)$ so $\sqrt{-1}\bar\partial\partial\phi = \Omega$.

2. Let g be a Kähler metric on $B_{\vec{\varepsilon}}(\vec{z}_0)$. Then g arises from a positive line bundle.

Proof. Let Ω be a real two-form of Type (1,1) on $B_{\vec{\varepsilon}}(\vec{z}_0)$ with $d\Omega = 0$. By the Poincaré Lemma (see Lemma 19.7), there is a smooth real one-form so $d\omega = \Omega$. Decompose

$$\omega = \omega^{1,0} + \omega^{0,1}$$

as the sum of forms of Type (1,0) and Type (0,1), respectively. As ω is real, $\overline{\omega^{1,0}} = \omega^{0,1}$. Because $d\omega = \Omega$ is of Type (1,1), $\partial\omega^{1,0} = 0$ and $\bar\partial\omega^{0,1} = 0$. The Dolbeault Lemma (see Theorem 19.7) shows there is a smooth complex function ϕ so that $\omega^{0,1} = \bar\partial\phi$ and, consequently, that $\omega^{1,0} = \partial\bar\phi$. Let $\psi := 2\Im(\phi) = -\sqrt{-1}(\phi - \bar\phi)$. We establish Assertion 1 by expressing

$$\Omega = d\omega = \bar\partial\omega^{1,0} + \partial\omega^{0,1} = \bar\partial\partial\bar\phi + \partial\bar\partial\phi = \bar\partial\partial(\phi - \bar\phi) = \sqrt{-1}\bar\partial\partial\psi \, .$$

Let Ω be the Kähler form associated to a Kähler metric on $B_{\vec{\varepsilon}}(\vec{z}_0) \times \mathbb{C}$. Define a fiber metric on the trivial line bundle $\mathbb{1} = B_{\vec{\varepsilon}}(\vec{z}_0) \times \mathbb{C}$ by setting $h = e^{2\pi\psi}$. We establish Assertion 2 by computing $c_1(\mathbb{1}, h) = \frac{\sqrt{-1}}{2\pi}\bar\partial\partial\log(h) = \sqrt{-1}\bar\partial\partial\psi = \Omega$. □

If g is a Kähler metric on M, then Ω_g is a closed real form of type (1, 1) and $[\Omega]$ determines an element of $H^2_{\mathrm{deR}}(M)$. This cohomology class plays an important role in the analysis of Kähler geometry. We have the following representation result.

Theorem 19.18   Let M be a simply connected compact holomorphic manifold. Let g_i be two Kähler metrics on M with associated Kähler forms Ω_i. If $[\Omega_1] = [\Omega_2]$ in de Rham cohomology, then there exists a smooth real-valued function ψ so that

$$\Omega_2 = \Omega_1 + \sqrt{-1}\bar\partial\partial\psi.$$

Proof. We use the same argument that we used to prove Lemma 19.17 to establish this result. Let $\Omega = \Omega_1 - \Omega_2$. By hypothesis, $[\Omega] = 0$ in $H^2_{\mathrm{deR}}(M)$. Consequently, there exists a real one-form ω so $d\omega = \Omega$. Decompose $\omega = \omega^{1,0} + \omega^{0,1}$. Since M is simply connected, $H^1_{\mathrm{deR}}(M) = 0$. Since M is Kähler, we may use Theorem 18.39 to see $H^{1,0}(M) = 0$ and $H^{0,1}(M) = 0$. Since $d\omega = \Omega$ is of Type $(1,1)$, we have $\bar{\partial}\omega^{0,1} = 0$. Since $H^{0,1}(M) = 0$, we can find a smooth complex function ϕ so $\omega^{0,1} = \bar{\partial}\phi$ and hence $\overline{\omega^{0,1}} = \partial\bar{\phi}$. Let ψ be the imaginary part of ϕ. We complete the proof by computing

$$\Omega = d\omega = \bar{\partial}\omega^{1,0} + \partial\omega^{0,1} = \bar{\partial}\partial\bar{\phi} + \partial\bar{\partial}\phi = \partial\bar{\partial}(\phi - \bar{\phi}) = \sqrt{-1}\partial\bar{\partial}\psi. \qquad\square$$

The function ϕ is called the *Kähler potential*. In general, $\sqrt{-1}\partial\bar{\partial}\psi$ is a real form of Type $(1,1)$. Let g_ψ be the associated symmetric bilinear form; $g_2 = g_1 + g_\psi$. This construction parametrizes all Kähler metrics once the de Rham cohomology class has been fixed. Conversely, of course, by taking ψ small in the C^2 topology, we can construct many Kähler metrics. If g_1 is Hodge, then $\Omega_{g_1} = c_1(L, h)$. Replacing h by $e^\tau h$ where τ is a suitable multiple of $\log(\psi + C)$ then shows g_2 is Hodge as well. Thus, as we saw in the proof of Lemma 19.16, an obstruction to a Kähler metric g being Hodge lies in whether $[\Omega_g] \in H^2(M; \mathbb{Z})$.

19.4   THE KODAIRA EMBEDDING THEOREM

By Lemma 19.14, any compact holomorphic submanifold of complex projective space is a Hodge manifold. It will be the focus of this section to establish the converse; any compact Hodge manifold admits a holomorphic embedding in complex projective space. This is the *Kodaira Embedding Theorem* [43]. It then follows by *Chow's Theorem* [17] that such a manifold is a smooth projective variety although we will not show this. We refer to Griffiths and Harris [31] or Huybrechts [40] for further details concerning the material of this section; notes by Mauri [49] are also useful. In Section 19.4.1, we define the blowup $\widetilde{M}$ and show it is independent of the particular choice of holomorphic coordinates. In Section 19.4.2, we examine certain geometric properties of the blowup. Let L be a positive line bundle over M. If $d(P_1, P_2) \geq \varepsilon > 0$, we will show in Section 19.4.3 that there exists $k(\varepsilon)$ so that if $k \geq k(\varepsilon)$, then there exist $s_i \in \mathrm{Hol}(L^k)$ so that $s_1(P_1) \neq 0$, $s_1(P_2) = 0$, $s_2(P_1) = 0$, and $s_2(P_2) \neq 0$. We will also show that there exists k_0 so that if $k \geq k_0$, then the exterior derivative d is a surjective map from $\mathrm{Hol}(L^k)$ to $\{L^k \otimes T^{1,0}\}(P)$. We will use these two results in Section 19.4.4 to establish the Kodaira Embedding Theorem.

19.4.1 THE BLOWUP. Let L be a holomorphic line bundle. The following sheafs will play an important role in the proof of the Kodaira Embedding Theorem. To clarify the notation, we will use boldface for sheafs in Section 19.4.

1. Let $\mathbf{Hol(L)}$ be the sheaf of holomorphic sections to L.

2. Let $\boldsymbol{L}_P$ be the sheaf associating to any open set containing P, the fiber of L at P. This sheaf is often called the *sky-scraper sheaf*.

3. The restriction map r_P^L maps the sheaf **Hol(L)** to the sheaf L_P.

4. Let $\{r_P^L\}$ be the sheaf of holomorphic sections to L vanishing at P.

Suppose $m > 1$. Lemma 19.8 shows that if $\phi \in \mathrm{Hol}(B_r(0) - \{0\})$, then ϕ extends holomorphically to $B_r(0)$. By considering the function ϕ^{-1}, we conclude there does not exist $\phi \in \mathrm{Hol}(B_r(0))$ only vanishing at the origin. Thus, there is no holomorphic line bundle which admits a holomorphic section which only vanishes at a single point. Consequently, $\{r_P^L\}$ is not the sheaf of holomorphic sections to any holomorphic line bundle and, therefore, the Kodaira Vanishing Theorem does not apply (directly) to this sheaf. To get around this difficulty, we introduce the blowup. We will construct a manifold $\widetilde{M}_P$ and a holomorphic projection $\pi : \widetilde{M}_P \to M$ so that $\pi^*\{r_P^L\}$ is the sheaf of holomorphic sections to a holomorphic line bundle over $\widetilde{M}_P$ (see Assertion 4 of Lemma 19.21). We proceed as follows to construct $\widetilde{M}_P$.

Definition 19.19   The line bundle $\mathbb{L} := \{((\langle \vec{w} \rangle), \vec{z}) \in \mathbb{CP}^{m-1} \times \mathbb{C}^m : \vec{z} \in \langle \vec{w} \rangle\}$ comes equipped with natural projections $\pi_1 : \mathbb{L} \to \mathbb{CP}^{m-1}$ and $\pi_2 : \mathbb{L} \to \mathbb{C}^m$. We may identify $\mathbb{CP}^{m-1}$ with the zero section $\pi_2^{-1}(0) = \mathbb{CP}^{m-1} \times \{0\}$ of $\mathbb{L}$. Let ν be the normal bundle of $\mathbb{CP}^{m-1} \times \{0\}$ in $\mathbb{L}$. Then $\nu = \pi_1^*\mathbb{L}$. Let P be a point of holomorphic manifold M of complex dimension m. Let $(z^1, \ldots, z^m)$ be local holomorphic coordinates on an open neighborhood $\mathcal{O}$ of P which are centered at P. Define

$$\widetilde{M}_P^{\mathcal{O}} := (M - \{P\}) \cup \mathbb{L} / \sim$$

where we identify $\mathcal{O} - \{P\}$ with $\pi_2^{-1}(\mathcal{O} - \{0\})$. What we have done is replace P by directions pointing away from P, i.e., by $\mathbb{P}(T_{c,P}(M)) = \mathbb{CP}^{m-1}$. We cut out P and glue in a copy of $\mathbb{CP}^{m-1}$; π_2 defines a natural holomorphic map $\pi_2 : \widetilde{M}_P^{\mathcal{O}} \to M$ which is a bi-holomorphic map from $\widetilde{M}_P^{\mathcal{O}} - \mathbb{CP}^{m-1}$ to $M - \{P\}$.

The manifold $\tilde{M}_P^{\mathcal{O}}$ depends on the choice of holomorphic coordinates. But we may show that changing $\mathcal{O}$ replaces $\tilde{M}_P^{\mathcal{O}}$ by a holomorphically equivalent manifold as follows. We can assume $M = \mathcal{O} \subset \mathbb{C}^m$. Changing the coordinates is equivalent to taking a bi-holomorphic map $\Phi : (\mathcal{O}, 0) \to (\mathcal{U}, 0)$ from one neighborhood of 0 to another neighborhood of 0 where $\Phi(0) = 0$. Define an extension Ψ from the blowup defined by $\mathcal{O}$ to the blowup defined by $\mathcal{U}$ by setting

$$\Psi(\langle z \rangle, w) = \left\{ \begin{array}{ll} (\langle \Phi w \rangle, \Phi w) & \text{if } w \neq 0 \\ (\langle d\Phi(0)z \rangle, 0) & \text{if } w = 0 \end{array} \right\}.$$

This is well defined as a set-theoretic map and it is clear $\pi_2 \circ \Psi = \Phi$. It is clearly holomorphic if $w \neq 0$ and it is clearly holomorphic on $\mathbb{CP}^{m-1}$. And it clearly is 1-1 and onto. So the only point at issue is how the two definitions fit together, i.e., is Ψ a holomorphic map?

Lemma 19.20   Adopt the notation established above.

1. Let X and Y be metric spaces. Let $f : X \to Y$. If given any sequence $x_n \to x_\infty$, there exists a subsequence x_{n_k} so $f(x_{n_k}) \to f(x_\infty)$, then f is continuous at x_∞.

2. If $0 \neq \xi_n \in \mathbb{C}^m$ is a sequence with $\lim_{n\to\infty} \xi_n = 0$ and $\lim_{n\to\infty} \langle \xi_n \rangle = \langle \xi_\infty \rangle$, then

$$\lim_{n\to\infty} \langle \Phi \xi_n \rangle = \langle d\Phi(0)\xi_\infty \rangle \, .$$

3. Ψ is continuous.

4. Ψ is a bi-holomorphic map.

5. The blowup $\tilde{M}_P := \tilde{M}_P^{\mathcal{O}}$ is independent of the holomorphic coordinate system chosen.

Proof. We argue by contradiction to prove the first assertion. Suppose f satisfies the hypothesis of the first assertion but, to the contrary, f is not continuous at x_0. Then there must exist $\varepsilon > 0$ so that for every $\delta > 0$ there exists x_δ with $d(x_\delta, x_\infty) < \delta$ but $d(f(x_\delta), f(x_\infty)) \geq \varepsilon$. Choose points x_n so $d(x_n, x_\infty) < \frac{1}{n}$ but $d(f(x_n), f(x_\infty)) \geq \varepsilon$. Since $x_n \to x_\infty$, we can choose a subsequence $x_{n_k} \to x_\infty$ so $f(x_{n_k}) \to f(x_\infty)$. This contradicts the fact that $d(f(x_{n_k}), f(x_\infty)) \geq \varepsilon > 0$ for all k and establishes the desired conclusion.

We now establish the second assertion. Suppose $0 \neq \xi_n \in \mathbb{C}^m$ satisfies $\xi_n \to 0$ and $\lim_{n\to\infty} \langle \xi_n \rangle = \langle \xi_\infty \rangle$. By passing to a subsequence if needed, we can assume the unit vectors $\|\xi_n\|^{-1} \xi_n \to \eta$ for some $\eta \in S^{2m-1}$. We then have $\langle \eta \rangle = \langle \xi_\infty \rangle$. Since Φ is complex differentiable and $\Phi(0) = 0$, we can write $\Phi\xi = d\Phi(0)\xi + O(\|\xi\|^2)$. Thus, since $d\Phi(0)$ is a continuous linear operator, we may establish the second assertion by computing:

$$\|\xi_n\|^{-1}\Phi(\xi_n) = d\Phi(0)\{\|\xi_n\|^{-1}\xi_n\} + \|\xi_n\|^{-1}O(\|\xi_n\|^2),$$
$$\lim_{n\to\infty} \|\xi_n\|^{-1}\Phi(\xi_n) = d\Phi(0)\eta,$$
$$\lim_{n\to\infty} \langle \Phi\xi_n \rangle = \lim_{n\to\infty} \langle \|\xi_n\|^{-1}\Phi(\xi_n) \rangle = \langle d\Phi(0)\eta \rangle = \langle d\Phi(0)\xi_\infty \rangle \, .$$

We use Assertion 2 to show that Ψ is continuous. Suppose we are given a sequence of points $(\langle P_n \rangle, \xi_n) \to (\langle P_\infty \rangle, \xi_\infty)$. We must show there exists a subsequence so that

$$\Psi(\langle P_n \rangle, \xi_n) \to \Psi(\langle P_\infty \rangle, \xi_\infty) \, . \qquad (19.4.a)$$

By passing to subsequences, it suffices to consider one of the following three cases.

1. All the $\xi_n = 0$. This implies $\xi_\infty = 0$ as well. In this case, since $d\Phi(0)$ is a linear map, $\langle P_n \rangle \to \langle P_\infty \rangle$ implies that $\langle d\Phi_n(0)P_n \rangle \to \langle d\Phi_n(0)P \rangle$ so Equation (19.4.a) holds.

2. None of the $\xi_n = 0$ and $\xi_\infty \neq 0$. Thus, we may set $P_n = \xi_n$ and $P_\infty = \xi_\infty$. Since Φ is continuous, Equation (19.4.a) holds.

3. None of the $\xi_n = 0$ but $\xi_\infty = 0$. This is the only non-trivial case. We may set $P_n = \xi_n$ and $\eta = P_\infty$. We apply Assertion 2 to see $\langle \Phi \xi_n \rangle \to \langle d\Phi(0)\eta \rangle$ and obtain Equation (19.4.a).

This completes the proof of Assertion 3. By Assertion 3, Ψ is continuous and, hence, locally bounded. It is holomorphic on $L - \mathbb{CP}^{m-1}$. We may then use Lemma 19.8 to see Ψ is holomorphic. $\qquad \square$

19.4.2 GEOMETRIC PROPERTIES OF THE BLOWUP. We establish certain properties of the blowup as follows. Let $\pi : \widetilde{M}_P \to M$ be the natural projection. In particular, we show $\pi^*\{r_P^L\}$ is the sheaf of holomorphic sections to a suitable line bundle over $\widetilde{M}_P$. Let π_1 and π_2 be as given in Definition 19.19. We have π_1 maps a neighborhood of the exceptional fiber $\mathbb{CP}^{m-1}$ to $\mathbb{CP}^{m-1}$. Since $\pi_1^*\mathbb{L}$ admits a canonical trivialization defined by π_2 away from $\mathbb{CP}^{m-1}$, we may regard $\pi_1^*\mathbb{L}$ as defined on all of $\widetilde{M}_P$.

Lemma 19.21

1. $\Lambda^{m,0}(\widetilde{M}_P) = \pi^*(\Lambda^{m,0}M) \otimes \pi_1^*(\mathbb{L})^{m-1}$.
2. If L is a holomorphic line bundle over M, $\pi^* : \mathrm{Hol}(L) \to \mathrm{Hol}(\pi^*L)$ is an isomorphism.
3. Let (L, h) be a positive line bundle over M. Given ℓ, there exists $k = k(\ell)$, which is independent of P, so that $(\pi^*L)^k \otimes \pi_1^*(\mathbb{L}^*)^\ell$ is a positive line bundle over $\widetilde{M}_P$.
4. $\pi_1^*\mathbb{L}$ is the normal bundle of $\mathbb{CP}^{m-1}$ in $\widetilde{M}_P$ and $\pi^* \ker(r_P^L) = \mathbf{Hol}(\pi_1^*\mathbb{L}^*)$.

Proof. We apply Lemma 19.9 to prove Assertion 1. Replace $\mathbb{CP}^m$ by $\mathbb{CP}^{m-1}$ to see

$$\Lambda^{m,0}\widetilde{M}_P = \pi_1^*(\mathbb{L})^{m-1} \quad \text{near} \quad \pi^{-1}(P) = \mathbb{CP}^{m-1}. \tag{19.4.b}$$

On the other hand, it is immediate from the definition that

$$\Lambda^{m,0}\widetilde{M}_P = \pi^*(\Lambda^{m,0}M) \quad \text{away from} \quad \pi^{-1}(P). \tag{19.4.c}$$

Note that π is a holomorphic diffeomorphism from $\widetilde{M}_P - \mathbb{CP}^{m-1}$ to $M - \{P\}$. The bundle $\pi_1^*\mathbb{L}$ is trivial on $\mathbb{L} - \mathbb{CP}^{m-1}$ since it admits a global section. The bundle $\Lambda^{m,0}M$ is trivial near P. Thus, we can glue together the isomorphisms of Equations (19.4.b) and (19.4.c) to derive Assertion 1.

Let $s \in \mathrm{Hol}(L)$. Since π is surjective, $\pi^*s = 0$ implies that $s = 0$. Conversely, let $\tilde{s}$ belong to $\mathrm{Hol}(\pi^*L)$ be given. Since π is a biholomorphic map from $\widetilde{M}_P - \mathbb{CP}^{m-1}$ to $M - \{P\}$, there exists $s \in \mathrm{Hol}(L|_{M-\{P\}})$ so $\pi^*s = \tilde{s}$ on $\widetilde{M}_P - \mathbb{CP}^{m-1}$. By Lemma 19.8, s extends holomorphically to a section to L over all of M. We then have by continuity that $\tilde{s} = \pi^*s$ on all of $\widetilde{M}_P$. This verifies Assertion 2.

Let $h_{\mathbb{L}^*}$ be the canonical fiber metric on $\mathbb{L}^*$ whose first Chern form determines the Fubini–Study metric. Let $\pi_1 : \mathbb{L}^* \to \mathbb{CP}^{m-1}$. Then the first Chern form of $\pi_1^*(\mathbb{L}^*, h_{\mathbb{L}^*})$ is positive on directions that are tangential to $\mathbb{CP}^{m-1}$ and vanishes on normal directions. We have $\pi_1^*(\mathbb{L}^*)$ is trivial on $\mathbb{L} - \mathbb{CP}^{m-1}$. Thus, using a cut off function, we can put a fiber metric on $\pi_1^*(\mathbb{L}^*)$ to agree with $\pi_1^*h_{\mathbb{L}^*}$ near $\mathbb{CP}^{m-1}$ and to be flat away from some small neighborhood of $\mathbb{CP}^{m-1}$. Since L is assumed positive, π_1^*L will be positive away from $\mathbb{CP}^{m-1}$, positive on normal directions to $\mathbb{CP}^{m-1}$ near $\mathbb{CP}^{m-1}$, and always non-negative. Taking high powers of L will counteract any negativity introduced by the cut-off function away from $\mathbb{CP}^{m-1}$; near $\mathbb{CP}^{m-1}$, the tangential directions are controlled by $\mathbb{L}^*$ and the normal directions are controlled by powers of L. This proves Assertion 3.

We noted previously that $\pi_1^*\mathbb{L}$ is the normal bundle of $\mathbb{CP}^{m-1}$ in $\widetilde{M}_P$. Thus, $\pi_1^*\mathbb{L}^*$ provides local holomorphic coordinates for the fibers of the normal bundle near $\mathbb{CP}^{m-1}$ which vanish on $\mathbb{CP}^{m-1}$; Assertion 4 now follows.   □

19.4.3 SECTIONS TO POSITIVE LINE BUNDLES. Let L be a holomorphic line bundle over a holomorphic manifold (M, J). If $P \in M$, let L_P be the fiber of L over P. If s is a holomorphic section defined near P, we let $r_P^L(s) := s(P) \in L_P$. If $s(P) = 0$, then $\partial s(P)$ is well defined and belongs to $L_P \otimes T_P^{1,0}(M)$.

Lemma 19.22   Let (L, h) be a positive line bundle over (M, J). Let $\varepsilon > 0$ be given. There exists an integer $k_0 = k_0(\varepsilon, L, h, M, J) > 0$ so that if $k \geq k_0$, then:

1. If $d(P_1, P_2) \geq \varepsilon > 0$, then $r_{P_1}^{L_{P_1}^k} \oplus r_{P_2}^{L_{P_2}^k} : \mathrm{Hol}(L^k) \to L_{P_1}^k \oplus L_{P_2}^k$ is surjective.

2. $r_P^{L^k} : \mathrm{Hol}(L^k) \to L_P^k$ is surjective.

3. $\partial_P : \{s \in \mathrm{Hol}(L^k) : s(P) = 0\} \to L_P^k \otimes T_P^{1,0}(M)$ is surjective.

Proof. Our proof of these assertions will be sheaf-theoretic in nature. We have a short exact sequence of sheaves (see Definition 16.8)

$$0 \to \ker(r_{P1}^L \oplus r_{P2}^L) \to \mathbf{Hol}(L) \xrightarrow{r_{P_1}^L \oplus r_{P_2}^L} L_{P_1} \oplus L_{P_2} \to 0$$

which, by Lemma 16.9, gives rise to a long exact sequence in cohomology

$$\begin{aligned}0 \to \ & H^0(M; \ker(r_{P1}^L \oplus r_{P2}^L)) \ \to \ H^0(M; \mathbf{Hol}(L)) \xrightarrow{r_{P_1}^L \oplus r_{P_2}^L} H^0(M; L_{P_1} \oplus L_{P_2}) \\ \to \ & H^1(M; \ker(r_{P1}^L \oplus r_{P2}^L)) \ \to \ H^1(M; \mathbf{Hol}(L)) \ \cdots .\end{aligned}$$

Identifying $H^0(M; \mathbf{Hol}(L))$ with the space of global holomorphic sections to L and $H^0(M; L_P)$ with the stalk L_P then yields

$$\cdots \mathrm{Hol}(L) \xrightarrow{r_{P_1}^L \oplus r_{P_2}^L} L_{P_1} \oplus L_{P_2} \to H^1(M; \ker(r_{P1}^L \oplus r_{P2}^L)) \cdots .$$

Thus, Assertion 1 would follow if we could show

$$H^1(M; \ker(r_{P1}^L \oplus r_{P2}^L)) = 0.$$

Let $\widetilde{M} := \widetilde{M}_{P_1, P_2}$ be the blowup of M at P_1 and P_2. Let $\widetilde{M}_i = \pi^{-1}(P_i)$ be the exceptional $\mathbb{CP}^{m-1}$ fibers. We have a commutative diagram of short exact sequences of sheaves

Diagram 19.4.1

$$\begin{array}{ccccccccc} 0 \to & \ker(r_{M_1}^{\pi^*L} \oplus r_{M_2}^{\pi^*L}) & \to & \pi^*L & \to & \pi^*L|_{M_1} \oplus \pi^*L|_{M_2} & \to 0 \\ & \uparrow \pi^* & \circ & \uparrow \pi^* & \circ & \uparrow \pi^* & \\ 0 \to & \ker(r_{P1}^L \oplus r_{P2}^L) & \to & L & \to & L|_{P_1} \oplus L|_{P_2} & \to 0 \end{array}.$$

We focus on the three terms of interest in the commutative diagram arising from the long exact sequences in cohomology arising from Diagram 19.4.1 to obtain

Diagram 19.4.2

$$
\begin{array}{ccccc}
H^0(\widetilde{M};\pi^*L) & \to & H^0(\widetilde{M};\pi^*L|_{M_1} \oplus \pi^*L|_{M_2}) & \to & H^1(\widetilde{M};\ker(r_{M_1}^{\pi^*L} \oplus r_{M_2}^{\pi^*L})) \\
\uparrow \pi^* & \circ & \uparrow \pi^* & \circ & \uparrow \pi^* \\
H^0(M;L) & \to & H^0(M;L|_{P_1} \oplus L|_{P_2}) & \to & H^1(M;\ker(r_{P_1}^L \oplus r_{P_2}^L))
\end{array}.
$$

We have $H^0(\widetilde{M};\pi^*L) = \mathrm{Hol}(\pi^*L)$ and $H^0(M;L) = \mathrm{Hol}(L)$. Lemma 19.21 shows that $\pi^* : \mathrm{Hol}(L) \to \mathrm{Hol}(\pi^*L)$ is an isomorphism. We have $H^0(M;\pi^*L|_{M_1 \cup M_2})$ is the space of holomorphic sections to π^*L over $M_1 \cup M_2$. The bundle π^*L is trivial over the union $M_1 \cup M_2$. Because $M_1 \cup M_2$ is compact, π^* is an isomorphism from $H^0(M;L|_{P_1 \cup P_2})$ to $H^0(\widetilde{M};\pi^*L|_{M_1 \cup M_2})$. Let $\mathbb{L}_i := \pi_{1,i}^* \mathbb{L}$. Then $\mathbb{L}_1 \otimes \mathbb{L}_2$ does admit a section which vanishes to first order on $M_1 \cup M_2$. Thus, $\ker\{r_{M_1}^{\pi^*L} \oplus r_{M_2}^{\pi^*L}\} = \mathrm{Hol}(\pi_{2,1}^* \mathbb{L} \otimes \pi_{2,2}^* \mathbb{L} \otimes \pi^*L)$. We replace L by L^k and set $V_k := \pi_{2,1}^* \mathbb{L} \otimes \pi_{2,2}^* \mathbb{L} \otimes \pi^*L^k$ to rewrite Diagram 19.4.2 as

Diagram 19.4.3

$$
\begin{array}{ccccc}
\mathrm{Hol}(\pi^*L^k) & \to & \mathrm{Hol}(\pi^*L^k|_{M_1}) \oplus \mathrm{Hol}(\pi^*L^k|_{M_2}) & \to & H^1(\widetilde{M};\mathbf{Hol}(V_k)) \\
\uparrow \pi^* \approx & \circ & \uparrow \pi^* \approx & \circ & \uparrow \pi^* \\
\mathrm{Hol}(L^k) & \to & L^k|_{P_1} \oplus L^k|_{P_2} & \to & H^1(M;\ker(r_{P_1}^{L^k} \oplus r_{P_2}^{L^k}))
\end{array}.
$$

By Assertion 1 of Lemma 19.21,

$$
V_k \otimes \Lambda^{m,0}\widetilde{M}^* = \pi^*(\Lambda^{m,0}M^*) \otimes \pi_{2,1}^*(\mathbb{L}^*)^{m-1} \otimes \pi_{2,2}^*(\mathbb{L}^*)^{m-1} \otimes \pi_{2,1}^*\mathbb{L}^* \otimes \pi_{2,2}^*\mathbb{L}^* \otimes \pi^*L^k .
$$

By Assertion 2 of Lemma 19.13, there exists k_0 so $L^k \otimes \Lambda^{m,0}M^*$ is positive for any $k \geq k_0$. Thus, $\pi^*(L^k \otimes \Lambda^{m,0}M^*)$ is positive semi-definite. By Assertion 3 of Lemma 19.21, choose k_1 so that $(\pi_{2,1}^*\mathbb{L}^*)^m \otimes L^{k_1}$ is positive and so that $(\pi_{2,1}^*\mathbb{L}^*)^m \otimes L^{k_1}$ is positive for $k \geq k_1$. This implies $V_k \otimes \Lambda^{m,0}\widetilde{M}^*$ is positive for $k \geq \max(k_0, k_1)$. We then use Lemma 19.13 to see $H^1(\widetilde{M};\mathbf{Hol}(V_k)) = 0$ and Assertion 1 follows from Diagram 19.4.3. The assumption that $d(P_1, P_2) \geq \varepsilon > 0$ is necessary since we are applying the construction used to prove Assertion 3 of Lemma 19.21 and we do not want the cut-off functions centered at P_1 and P_2 to interact. Assertion 2 is an immediate consequence of Assertion 1. The proof of Assertion 3 is similar. The map d_P arises from the short exact sequence of sheafs

$$
0 \to \ker(d_P) \to \ker(r_P^L) \xrightarrow{d_P} L_P \otimes T_P^{1,0}(M) \to 0 .
$$

We may identify

$$
\pi^*\ker(d_P) = \pi_1^*(\mathbb{L}^*)^2 \otimes \pi^*L \quad \text{and} \quad \ker(r_P^L) = \pi_1^*(\mathbb{L}^*) \otimes \pi^*L .
$$

Applying exactly the same argument as that used to prove Assertion 1 then yields Assertion 3. □

19.4.4 THE KODAIRA EMBEDDING THEOREM.

Theorem 19.23   If M is a compact Hodge manifold, then there exists a holomorphic embedding of M in $\mathbb{CP}^n$ for some n.

Proof. Let (L, h) be a positive line bundle over M. We apply Lemma 19.22. By replacing (L, h) by (L^k, h^k), we may assume $\partial_P : \{s \in \mathrm{Hol}(L) : s(P) = 0\} \to L_P \otimes T_P^{1,0}(M)$ is surjective and that $r_P^L : \mathrm{Hol}(L) \to L_P$ is surjective for any $P \in M$. Let $\{s^0, \ldots, s^n\}$ be a basis for $\mathrm{Hol}(L)$. Given P, there is some s^i so $s^i(P) \neq 0$. Then if $d(P, Q) < \delta_P$, $s^i(Q) \neq 0$. We may express $s^j(Q) = \phi_i^j(Q)s^i(Q)$ where the functions ϕ_i^j are holomorphic. We let

$$\Phi_i := (\phi_i^0, \ldots, \phi_i^n) : B_\delta(P) \to \mathbb{C}^{n+1} - \{0\}.$$

Let $\langle \Phi_i \rangle$ be the corresponding holomorphic map from $B_{\delta_P}(P)$ to $\mathbb{CP}^n$. If $s_j(P) \neq 0$, then $\Phi_i = \phi_j^i \Phi_j$ so $\langle \Phi_i \rangle = \langle \Phi_j \rangle$ is independent of the particular index chosen and we obtain a holomorphic map $\Phi : M \to \mathbb{CP}^n$. Since ∂_P is surjective, it follows that Φ is an embedding and, by compactness, there exists $\varepsilon > 0$ so that if $d(P, Q) < \varepsilon$, then there exists a section s_P to L so $s_P(P) = 0$ and $s_P(Q) \neq 0$. Fix this ε. Apply Lemma 19.22. If $d(P, Q) \geq \varepsilon$, there exists a section $t_{P,Q}$ to L^ℓ so $t_{P,Q}(P) = 0$ and $t_{P,Q}(P) \neq 0$. If $d(P, Q) \leq \varepsilon$, then $s_P^\ell(P) = 0$ and $s_P^\ell(Q) \neq 0$. Thus, the associated map ϕ_ℓ is 1-1 and an embedding. □

Bibliography

[1] Y. Akizuki and S. Nakano, Note on Kodaira–Spencer's proof of Lefschetz theorems, *Proceedings of the Japan Academy* **30** (1954), 266–272. DOI: 10.3792/pja/1195526105. 97

[2] B. Allard, Private communication circa 1974. 53

[3] C. Arzelà, Unósservazione intorno alle serie di funzioni, *Rend. Dell' Accad. R. Delle Sci. Dell'Istituto di Bologna* (1882–1883), 142–159. 11

[4] C. Arzelà, Sulle funzioni di linee, *Mem. Accad. Sci. Ist. Bologna Cl. Sci. Fis. Mat.* **5** (1895), 55–74.

[5] G. Ascoli, Le curve limite di una varietà data di curve, *Atti della R. Accad. Dei Lincei Memorie della Cl. Sci. Fis. Mat. Nat.* **18** (1883–1884), 521–586. 11

[6] M. F. Atiyah, R. Bott, and A. Shapiro, Clifford modules, *Topology* **3** Suppl. 1 (1964), 3–38. DOI: 10.1016/0040-9383(64)90003-5. 60

[7] R. Baire, *Sur les fonctions de variables réelles*, Thèses, Université de Paris Faculté des Sciences Mathématiques (1899), publ. Bernardoni de C. Rebeschini. DOI: 10.1007/bf02419243. 11

[8] W. Ballman, *Lectures on Kähler manifolds*, European Mathematical Society (2006). 92

[9] C. Birkenhake and H. Lange, *Complex tori*. Progress in Mathematics **177** (1999), Boston, MA, Birkhäuser Boston. DOI: 10.1007/978-1-4612-1566-0. 121

[10] S. Bochner, Vector fields and Ricci curvature, *Bull. Amer. Math. Soc.* **52** (1946), 776–797. DOI: 10.1090/s0002-9904-1946-08647-4. 69

[11] T. Branson and P. Gilkey, Residues of the eta function for an operator of Dirac type, *J. Functional Analysis* **108** (1992), 47–87. DOI: 10.1016/0022-1236(92)90146-a. 66

[12] V. Bunyakovsky, Sur quelques inegalités concernant les intégrales aux différences finies, *Mem. Acad. Sci. St. Petersbourg* **7** (1859), 1–18. 17

[13] J. W. Calkin, Two-Sided Ideals and Congruences in the Ring of Bounded Operators in Hilbert Space, *Annals of Math.* **42** (1941), 839–873. DOI: 10.2307/1968771. 27

[14] H. Cartan and S. Eilenberg, *Homological algebra*, Princeton University Press, Princeton, N. J. (1956). DOI: 10.1515/9781400883844. 7

[15] A. L. Cauchy, Cours d'analyse de l'École royale polytechnique. Première partie : Analyse algébrique Öuvres complètes, série 2, tome 3, 5-5+i-viii+17-471. (See Note II Theorem 16). See also English Translation Bradley, Robert E.; Sandifer, C. Edward *Cauchy's Cours d'analyse. An annotated translation*. Sources and Studies in the History of Mathematics and Physical Sciences. Springer, New York, 2009. xx+411 pp. (see page 303). DOI: 10.1007/978-1-4419-0549-9. 17

[16] S. Chern, A simple intrinsic proof of the Gauss-Bonnet formula for closed Riemannian manifolds, *Annals of Math.* (2) **45** (1944), 741–752. DOI: 10.2307/1969302. 69

[17] W. L. Chow, On Compact Complex Analytic Varieties, *Amer. J. Math.* **71** (1949), 893–914. DOI: 10.2307/2372375. 122

[18] J. Conway, *A course in functional analysis,* 2nd *ed.* Graduate Texts in Mathematics, 96. Springer-Verlag, New York, 1990. DOI: 10.1007/978-1-4757-3828-5. 16

[19] G. de Rham, La théorie des formes différentielles extérieures et l'homologie des variétés différentiables, *Rend. Mat. Appl.* **20** (1961), 105–146. DOI: 10.1007/978-3-642-10952-2_1. 69

[20] P. Dolbeault, Sur la cohomologie des variétés analytiques complexe, *C. R. Acad. Sci. Paris* **236**, 175–277. 105, 106

[21] J. Fourier, *Théorie analytique de la chaleur*, (1822) Paris: Firmin Didot, père et fils. DOI: 10.1017/CBO9780511693229. 33

[22] M. Fréchet, Sur les ensembles de fonctions et les opérations linéaire, *C. R. Acad. Sci. Paris* **144** (1907), 1411–1416. 23

[23] E. I. Fredholm, Sur une classe d'equations fonctionnelle, *Acta Math* **27** (1903), 365–390. DOI: 10.1007/bf02421317. 27

[24] G. Fubini, Sulle metriche definite da una forme Hermitiana, *Atti del Reale Istituto Veneto di Scienze, Lettere ed Arti* (1904), 63 pp., 502–513. 109

[25] L. Gårding, Dirichlete's problem for linear elliptic partial differential equations, *Math. Scand.* **1** (1953), 55–72. 51

[26] P. Gilkey, *Invariance theory, the heat equation, and the Atiyah-Singer index theorem.* Second edition. Studies in Advanced Mathematics, CRC Press, Boca Raton, FL, (1995). DOI: 10.1201/9780203749791. 8

[27] P. Gilkey, *Geometric Properties of Natural Operators Defined by the Riemann Curvature Tensor*. World Scientific. DOI: 10.1142/4812. 115

[28] P. Gilkey, *The Geometry of Spherical Space Form Groups*, Second edition, World Scientific Press (Singapore) (Series in Pure Mathematics: Volume 28) (2018). DOI: 10.1142/0868. 80

[29] E. Goursat, *Cours d'analyse mathématique*, Gauthier-Villars (Paris), 1910. 102

[30] J. P. Gram, Undersøgelser angaaende Maengden af Primtal under en given Graeense, *Det K. Videnskabernes Selskab.* **2** (1884), 183–308. 19

[31] P. Griffiths and J. Harris, *Principles of Algebraic Geometry*. Pure and Applied Mathematics (1978). Wiley-Interscience, John Wiley & Sons, New York. DOI: 10.1002/9781118032527. 5, 121, 122

[32] M. Gromov and H. B. Lawson, Spin and scalar curvature in the presence of a fundamental group, *Annals of Math.* **111** (1980), 209–230. DOI: 10.2307/1971198. 80, 81

[33] A. Grothendieck, Sur quelques points d'algèbre homologique, I, *Tohoku Math. J.* **9** (1957), 119–221. DOI: 10.2748/tmj/1178244839. 5

[34] R. C. Gunning, *Lectures on Riemann Surfaces, Jacobian Varieties*, Math. Notes, Princeton University Press, 1972. DOI: 10.1515/9781400872695. 121

[35] F. Hartogs, Einige Folgerungen aus der Cauchyschen Integralformel bei Funktionen mehrerer Vernderlichen, Sitzungsberichte der Königlich Bayerischen Akademie der Wissenschaften zu München, *MathematischPhysikalische Klasse* **36** (1906), 223–242. 106

[36] F. Hartogs, Zur Theorie der analytischen Funktionen mehrerer unabhängiger Veränderlichen, insbesondere über die Darstellung derselber durch Reihen welche nach Potentzen einer Veränderlichen fortschreiten, *Math. Ann.* (in German), **62** (1906), 1–88. DOI: 10.1007/BF01448415. 106

[37] R. Hartshorne, *Algebraic geometry*, Springer-Verlag, Berlin, 1977. DOI: 10.1007/978-1-4757-3849-0. 5

[38] F. Hirzebruch, *Topological methods in algebraic geometry*, Springer-Verlag (Berlin) 1966. DOI: 10.1007/978-3-642-62018-8. 10, 90

[39] W. V. D. Hodge, *The Theory and Applications of Harmonic Integrals*, Cambridge University Press (1941). 56, 69

[40] D. Huybrechts, *Complex Geometry*, Springer-Verlag, Berlin, 2005. DOI: 10.1007/b137952. 122

[41] M. Karoubi, *K-theory. An introduction*, Springer-Verlag, Berlin, 2008. DOI: 10.1007/978-3-540-79890-3. 9, 65

[42] K. Kodaira, On a differential-geometric method in the theory of analytic stacks, *Proc. Natl. Acad. Sci. USA*, **39** (1953), 1268–1273. DOI: 10.1073/pnas.39.12.1268. 97

[43] K. Kodaira, On Kähler varieties of restricted type (an intrinsic characterization of algebraic varieties), *Annals of Math.* **60** (1954), 28–48. DOI: 10.2307/1969701. 122

[44] P. Koebe, Über die Uniformisierung reeller analytischer Kurven, *Göttinger Nachrichten* (1907), 177–190. 100

[45] E. Kummer, Über die Flächen vierten Grades mit sechzehn singulären Punkten, *Berl. Monatsber* **6** (1864), 246–260. 118

[46] T. Levi-Civita, Nozione di parallelismo in una varietà qualunque e conseguente specificazione geometrica della curvatura riemanniana, *Rendiconti del Circolo Matematico di Palermo* **42** (1916), 173–214. DOI: 10.1007/BF03014898. 4

[47] A. Lichnerowicz, Spineurs harmoniques, *C. R. Acad. Sci. Paris* **257** (1963), 7–9. 77, 80

[48] Mathworld, http://mathworld.wolfram.com/KummerSurface.html. 118

[49] M. Mauri, *The Kodaira embedding theorem*, https://www.math.u-psud.fr/~thomine/fr/archives/Enseignement1415/SemiM2/Mauri.pdf. 122

[50] J. Milnor and J. Stasheff, *Characteristic Classes*, Princeton University Press (1974), 7

[51] S. Nakano, On complex analytic vector bundles, *J. Math. Society Japan* **7** (1955), 1–12. DOI: 10.2969/jmsj/00710001. 95, 96

[52] A. Newlander and L. Nirenberg, Complex analytic coordinates in almost complex manifolds, *Annals of Math.* **65** (1957), 391–404. DOI: 10.2307/1970051. 84, 85

[53] Y. Nikolayevsky, Osserman manifolds of dimension 8, *Manuscripta Math.* **115** (2004), 31–53. DOI: 10.1007/s00229-004-0480-y. 115

[54] Y. Nikolayevsky, Osserman conjecture in dimension $\neq 8, 16$, *Math. Ann.* **331** (2005), 505–522. DOI: 10.1007/s00208-004-0580-8.

[55] Y. Nikolayevsky, On Osserman manifolds of dimension 16, *Contemporary geometry and related topics*, Univ. Belgrade Fac. Math., Belgrade, (2006), 379–398.

[56] R. Osserman, Curvature in the eighties, *Amer. Math. Monthly* **97** (1990), 731–756. DOI: 10.1080/00029890.1990.11995659. 115

[57] O. Pekonen, The Einstein field equation in a multidimensional universe, *General relativity and Gravitation* **20** (1988), 667–670. DOI: 10.1007/BF00758971. 10

[58] M. Plancherel, Contribution à l'étude de la représentation d'une fonction arbitraire par les intégrales définies, *Rendiconti del Circolo Matematico di Palermo* **30** (1910), 289–335. DOI: 10.1007/BF03014877. 36

[59] H. Poincaré, Analysis situs, *Journal de l'École Polytechnique* **1** (1895), 1–123. 70, 105

[60] H. Poincaré, Sur l'uniformisation des fonctions analytiques, *Acta Math* **31** (1908), 1–63. DOI: 10.1007/bf02415442. 100

[61] D. Pompeiu, Sur la continuité des fonctions de variables complexes, *Annales de la Faculté des Sciences de Toulouse* **2** (1905), 265–315. 102

[62] F. Rellich, Ein Satz über mittlere Konvergenz, *Nachrichten Göttingen* (1930), 30–35. 41, 49

[63] B. Riemann, Theorie der Abel'schen Functionen, *Journal für die reine und angewandte Mathematik* **54** (1857), 101–155. 121

[64] F. Riesz, Sur les opérations fonctionnelles linéaires, *C. R. Acad. Sci. Paris* **144** (1907), 1409–1411. 23

[65] G. Roch, Über die Anzahl der willkurlichen Constanten in algebraischen Functionen, *Journal für die reine und angewandte Mathematik* **64** (1865), 372–376, DOI: 10.1515/crll.1865.64.372. 90

[66] H. A. Schwarz, Über ein Flächen kleinsten Flächeninhalts betreffendes Problem der Variationsrechnung, *Acta Societatis Scientiarum Fennicae* (1888), 318–361. 17

[67] J. P. Serre, Faisceaux analytiques sur l'espace projectif, *Sèminaire Henri Cartan*, 6 (Talk no. 18): 1–10. 106

[68] S. Sobolev, On a theorem of functional analysis, *Am. Math. Soc., Transl.*, II. Ser. 34 (1963), 39–68; translation from Mat. Sb., N. Ser. 4 (46) (1963), 471–497. DOI: 10.1090/trans2/034/02. 40, 50

[69] E. Stiefel, Über Richtungsfelder in den projektiven Räumen und einen Satz aus der reellen Algebra, *Comment. Math. Helv.* **13** (1941), 201–218. DOI: 10.1007/BF01378061. 7

[70] M. H. Stone, The generalized Weierstrass approximation theorem, *Mathematics Magazine* **21** (1948), 167–184. DOI: 10.2307/3029750. 13

[71] E. Study, Kürzeste Wege im komplexen Gebiet, *Math. Ann.* bf 60 (1904), 321–378. DOI: 10.1007/BF01457616. 109

[72] F. Treves, *Topological vector spaces, distributions and kernels*, Dover Publications, Inc., Mineola, NY (1967). 16

[73] C. von Clausewitz, *On War*. Translated by O.J. Matthijs Jolles, Modern Library. xv

[74] Y. Watanabe, On the characteristic function of harmonic Kählerian spaces, *Tohoku Math. J.* **27** (1975), 13–24. DOI: 10.2748/tmj/1178241030. 114

[75] K. Weierstrass, Mathematische Werke, Band 3, Abhandlungen III, pp. 1–37, especially p. 5 (Sitzungsberichte, Kon.Preussischen Akademie der Wissenschaften, July 9 and July 30, 1885). 13

[76] A. Weil, *Scientific works. Collected papers II (1951–1964)*, Springer-Verlag, New York-Heidelberg, 1979. 118

[77] H. Whitney, Differentiable manifolds, *Annals of Math.* **37** (1936), 645–680. DOI: 10.2307/1968482. 7, 21

[78] Wikipedia, https://en.wikipedia.org/wiki/K3_surface. 118

Authors' Biographies

Esteban Calviño-Louzao

E. García-Río

Peter B Gilkey

JeongHyeong Park

Ramón Vázquez-Lorenzo

Esteban Calviño-Louzao[1] is a member of the research group in Riemannian Geometry at the Department of Geometry and Topology of the University of Santiago de Compostela (Spain). He received his Ph.D. in 2011 from the University of Santiago under the direction of E. García-Río and R. Vázquez-Lorenzo. His research specialty is Riemannian and pseudo-Riemannian geometry. He has published more than 20 research articles and books.

Eduardo García-Río[2] is a Professor of Mathematics at the University of Santiago de Compostela (Spain). He is a member of the editorial board of *Differential Geometry and its Applications* and *The Journal of Geometric Analysis* and leads the research group in Riemannian Geometry at the Department of Geometry and Topology of the University of Santiago de Compostela (Spain). He received his Ph.D. in 1992 from the University of Santiago under the direction of A. Bonome

[1] Dir. Xeral de Educación, Formación Profesional e Innovación Educativa, San Caetano, s/n, 15781 Santiago de Compostela, Spain. *email:* estebcl@edu.xunta.es

[2] Department of Mathematics, Faculty of Mathematics, University of Santiago de Compostela, 15782 Santiago de Compostela, Spain. *email:* eduardo.garcia.rio@usc.es

and L. Hervella. His research specialty is Differential Geometry. He has published more than 120 research articles and books.

Peter B Gilkey[3] is a Professor of Mathematics. He is a fellow of the American Mathematical Society and is a member of the editorial board of *Differential Geometry and its Applications* and *The Journal of Geometric Analysis*. He received his Ph.D. in 1972 from Harvard University under the direction of L. Nirenberg. His research specialties are Differential Geometry, Elliptic Partial Differential Equations, Algebraic Topology, and Linguistics. He has published more than 285 research articles and books.

JeongHyeong Park[4] is a Professor of Mathematics at Sungkyunkwan University and is an associate member of the KIAS (Korea). She received her Ph.D. in 1990 from Kanazawa University in Japan under the direction of H. Kitahara. Her research specialities are Differential Geometry and Global Analysis. She organized the geometry section of AMC 2013 (The Asian Mathematical Conference 2013), the ICM 2014 satellite conference on Geometric analysis, and geometric structures on manifolds (2016). She has published more than 98 research papers and books.

Ramón Vázquez-Lorenzo[5] is a member of the research group in Riemannian Geometry at the Department of Geometry and Topology of the University of Santiago de Compostela (Spain). He is a member of the Spanish Research Network on Relativity and Gravitation. He received his Ph.D. in 1997 from the University of Santiago de Compostela under the direction of E. García-Río and R. Castro. His research focuses mainly on Differential Geometry with special emphasis on the study of the curvature and the algebraic properties of curvature operators in the Lorentzian and in the higher signature settings. He has published more than 60 research articles and books.

[3]Mathematics Department, University of Oregon, Eugene OR 97403 USA. *email:* `gilkey@uoregon.edu`
[4]Department of Mathematics, Sungkyunkwan University, Suwon, 16419, Korea. *email:* `parkj@skku.edu`
[5]Department of Geometry and Topology, Faculty of Mathematics, University of Santiago de Compostela, 15782 Santiago de Compostela, Spain. *email:* `ravazlor@edu.xunta.es`

Index

Printed in the United States
by Baker & Taylor Publisher Services